The Latin American Studies Book Series

The Latin American Studies Book Series promotes quality scientific research focusing on Latin American countries. The series accepts disciplinary and interdisciplinary titles related to geographical, environmental, cultural, economic, political and urban research dedicated to Latin America. The series publishes comprehensive monographs, edited volumes and textbooks refereed by a region or country expert specialized in Latin American studies.

The series aims to raise the profile of Latin American studies, showcasing important works developed focusing on the region. It is aimed at researchers, students, and everyone interested in Latin American topics.

Submit a proposal: Proposals for the series will be considered by the Series Advisory Board. A book proposal form can be obtained from the Publisher, Juliana Pitanguy (juliana.pitanguy@springer.com).

More information about this series at http://www.springer.com/series/15104

Hugo Inda Ferrero · Felipe García Rodríguez
Editors

Advances in Coastal Geoarchaeology in Latin America

Selected papers from the GEGAL Symposium at La Paloma, Uruguay

Springer

Editors
Hugo Inda Ferrero
Centro Universitario Regional del Este
Universidad de la República
Rocha, Uruguay

Felipe García Rodríguez
Centro Universitario Regional del Este
Universidad de la República
Rocha, Uruguay

ISSN 2366-3421 ISSN 2366-343X (electronic)
The Latin American Studies Book Series
ISBN 978-3-030-17830-7 ISBN 978-3-030-17828-4 (eBook)
https://doi.org/10.1007/978-3-030-17828-4

This Springer imprint is published by the registered company Springer Nature Switzerland AG
The registered company address is: Gewerbestrasse 11, 6330 Cham, Switzerland

Preface

The Latin American Group for Geoarchaeological Studies (GEGAL) was founded in 2012 during the VIth Meeting of Archaeological Theory in South America, held at Goiás, Brazil. At this time, a group of archaeologists and geologist agreed about the need for a multidisciplinary study group on geoarchaeological topics, aimed at the development of a platform for the exchange and integration of experiences, information, methods, and techniques across Latin America.

The initial optimistic effort was welcomed by a wide arrange of researchers from all over the continent and various disciplines, currently including archaeologists, geologists, geographers, geomorphologists, and historians from 13 Latin American countries. Soon after the establishment of the GEGAL, the group decided to held thematic workshops on a regular basis. The first workshop was celebrated at Olavarría, Buenos Aires, Argentina in May 2013 devoted to "Soils in Archaeology". Since then, workshops were celebrated on a yearly basis and hosted by different countries including Brazil, Chile, Colombia, and Uruguay.

The IVth Workshop was held at La Paloma beach, in the Atlantic coast of Rocha county, Uruguay during December 2016. The workshop was devoted to Coastal Geoarchaeology and aimed to contribute with the progress of multidisciplinary approaches including the following topics: stratigraphy, sedimentology, pedology, and formation processes of the archaeological record; paleoenvironments and human occupations; geomorphology and reconstruction of ancient landscapes; absolute and relative chronologies; and detection and preservation of archaeological sites. It is worth to mention that the Workshop provided the right environment to discuss several aspects of Geoarcheology, comprising current threats to coastal archaeological sites due to climate change, new methodological developments, the uncertainty posed by the ever-changing sociopolitical context in Latin America, heritage policies and science development and funding in our countries.

The Workshop developed oral and poster presentations in plenary sessions planned to favor the multidisciplinary analysis of the contributions. Participants from Argentina, Brazil, Chile, Colombia, Mexico, and Uruguay attended the event and presented their contributions that were evaluated by the Scientific Board of the

Organizing Committee and the abstracts were later published as digital media which was made available to all participants upon registration.

Because most of the participants have never been in the Atlantic coast of Uruguay before, a post Workshop field trip was organized to the Cabo Polonio Protected Area to expose the environments, landscapes, geological and geomorphological features as well as the archaeological record of such breathtaking place to the Workshop attendants. The granitic outcrops that formed the cape and the after-shore dune fields were visited and their features, together with prehistoric human occupations of the area were discussed in the field.

This volume presents 10 papers which were selected by their scientific quality and methodological diversity. As is expected in Geoarchaeology, these papers merged approaches, methods, and techniques from several disciplines, reflecting at the same time the multidisciplinary nature of our field and the diversity of the scientific community that attended the workshop.

Finally, it is clear that Geoarcheology is at the same time, one of the cornerstones of Archeological science by providing the context for multidisciplinary approaches and a growing field, as evidenced by more and more publications devoted to the subject in the region and around the globe. I do hope that the present contribution will be a kind of benchmark in the progress of Geoarchaeology in Latin America, showing the evolution and advancements of regional studies to a broader audience.

Rocha, Uruguay Hugo Inda Ferrero

Acknowledgements

I would like to deeply thank Prof. Jorge Rabassa for providing guidance and support since early stages of manuscript preparation and for their helpful suggestions. The Springer edition team is also acknowledged for their support and patience.

I gratefully acknowledge all the contributing colleagues for their enthusiasm and disposition through the edition process.

I wish to thank the colleagues from the Organizing Committee that worked hard to make this event possible and during the workshop to make every attendant to feel at home: Laura del Puerto, Irina Capdepont, and Alfonso Machado. Also, I wish to thank the authorities and staff of the Centro Cultural La Paloma that hosted the Workshop.

Contents

Contributors

Marco Alvarez FCNyM - División de Mineralogía, Petrología y Sedimentología - Facultad de Ciencias Naturales y Museo (UNLP), La Plata, Argentina

Rodrigo Costa Angrizani CONICET - Centro de Investigaciones Científicas y Transferencia de Tecnología a la Producción (UADER), Diamante, Entre Ríos, Argentina

Eduardo Apolinaire CONICET - Facultad de Ciencias Naturales y Museo (UNLP), La Plata, Argentina

Alfonso Machado Arnaud Facultad de Ciencias, Universidad de la República, Montevideo, Uruguay

Gustavo Piñeiro Barceló Facultad de Ciencias, Universidad de la República, Montevideo, Uruguay

Andreina Bazzino Equipo técnico Proyecto RLA0058-AIEA, Montevideo, Uruguay

Adriana Blasi CIC - División de Mineralogía Petrología y Sedimentología - Facultad de Ciencias Naturales y Museo (UNLP), La Plata, Argentina

Roberto Bracco Boksar Facultad de Humanidades y Ciencias de la Educación, Universidad de la República, Montevideo, Uruguay;
Laboratorio 14C, Facultad de Química, Universidad de la República - Ministerio de Educación y Cultura, Montevideo, Uruguay;
CURE, Sede Rocha, Universidad de la República, Rocha, Uruguay

Irina Capdepont Laboratorio de Estudios del Cuaternario (MEC-UNCIEP, IECA Facultad de Ciencias), Centro Universitario Regional del Este (CURE-Rocha), Universidad de la República, Montevideo, Uruguay

Carola Castiñeira Latorre CONICET- Museo Argentino de Ciencias Naturales "Bernardino Rivadavia", Buenos Aires, Argentina

Rosiclér Theodoro da Silva Pontifícia Universidade Católica de Goiás, Goiânia, GO, Brazil

Daniela De Hoz Programa de Arqueología Subacuática, Instituto Nacional de Antropología y Pensamiento Latinoamericano, Ciudad de Buenos Aires, Argentina

Julio Cezar Rubin de Rubin Pontifícia Universidade Católica de Goiás, Goiânia, GO, Brazil

Laura del Puerto Centro Universitario Regional del Este (CURE), Universidad de la República, Rocha, Uruguay

Matías Dourteau Programa de Arqueología Marítima y Subacuática, Centro de Investigaciones del Patrimonio Costero, Universidad de la República, Maldonado, Uruguay

Christopher Duarte Laboratorio OSL, UNCIEP, Instituto de Ecología y Ciencias Ambientales (IECA), Facultad de Ciencias, Universidad de la República, Montevideo, Uruguay

Dolores Elkin CONICET and Programa de Arqueología Subacuática, Instituto Nacional de Antropología y Pensamiento Latinoamericano, Ciudad de Buenos Aires, Argentina

Cristian M. Favier Dubois Facultad de Ciencias Sociales, Departamento de Arqueología, INCUAPA – CONICET, Universidad Nacional del Centro de la provincia de Buenos Aires (UNICEN), Olavarría, Argentina

Catarina Menezes Ferreira Universidade Federal de Sergipe, São Cristóvão, SE, Brazil

Gabriele Viega Garcia Instituto Homem Brasileiro, Cuiabá, MT, Brazil

Avto Gogichaishvili Laboratorio Interinstitucional de Magnetismo Natural, Instituto de Geofísica, Universidad Nacional Autónoma de México (UNAM), Unidad Michoacán, Campus Morelia, Morelia, Mexico

Ofelia Gutiérrez UNCIEP, Instituto de Ecología y Ciencias Ambientales (IECA), Facultad de Ciencias, Universidad de la República, Montevideo, Uruguay

Hugo Inda Ferrero Centro Universitario Regional del Este (CURE), Universidad de la República, Rocha, Uruguay

Eduardo Keldjian Programa de Arqueología Marítima y Subacuática, Centro de Investigaciones del Patrimonio Costero, Universidad de la República, Maldonado, Uruguay

Roberto Kokot Facultad de Filosofía y Letras, Universidad de Buenos Aires, Ciudad de Buenos Aires, Argentina

Laura Maravilla FCNyM - División de Mineralogía, Petrología y Sedimentología - Facultad de Ciencias Naturales y Museo (UNLP), La Plata, Argentina

Florencia Mari LATYR - Laboratorio de Radiocarbono, Centro de Investigaciones Geológicas. CONICET-UNLP, La Plata, Buenos Aires, Argentina

José López Mazz Facultad de Humanidades y Ciencias de la Educación, Universidad de la República, Montevideo, Uruguay

Juan Morales Laboratorio Interinstitucional de Magnetismo Natural, Instituto de Geofísica, Universidad Nacional Autónoma de México (UNAM), Unidad Michoacán, Campus Morelia, Morelia, Mexico

Rosana Najjar Instituto de Patrimônio Histórico e Artístico Nacional, Rio de Janeiro, RJ, Brazil

Rosario Odino Dirección Nacional de Minería y Geología (DINAMIGE), MIEM, Montevideo, Uruguay

Daniel Panario UNCIEP, Instituto de Ecología y Ciencias Ambientales (IECA), Facultad de Ciencias, Universidad de la República, Montevideo, Uruguay

Ethel Reina Dirección Nacional de Minería y Geología (DINAMIGE), MIEM, Montevideo, Uruguay

Leda Sánchez Bettucci Laboratorio de Geofísica y Geotectónica, Facultad de Ciencias, Departamento de Geodinámica Interna, Universidad de la República, Montevideo, Uruguay

Marcos Tassano Laboratorio de Radioquímica, Centro de Investigaciones Nucleares, Facultad de Ciencias, Universidad de la República, Montevideo, Uruguay

Rodrigo Torres Programa de Arqueología Marítima y Subacuática, Centro de Investigaciones del Patrimonio Costero, Universidad de la República, Maldonado, Uruguay

Dalila Viviana Herrera Villegas INCUAPA – CONICET, Facultad de Ciencias Sociales, Departamento de Arqueología, Universidad Nacional del Centro de la Provincia de Buenos Aires (UNICEN), Olavarría, Argentina

Michael Zech Landscape Ecology, Institute of Geography, Technical University of Dresden, Dresden, Germany

Human Occupation Chronologies Modeled by Geomorphological Factors: A Case Study from the Atlantic Coast of Northern Patagonia (Argentina)

Cristian M. Favier Dubois

Abstract Radiocarbon chronologies obtained in a dynamic landscape such as the Atlantic Ocean coast provide valuable information about the different stages of their use, but they also reflect inevitable biases, many of which are linked to geomorphological factors. The study of these factors and their changes over time becomes an indispensable task in which geoarchaeology plays an important role. In this paper, the Atlantic coast of Northern Patagonia (San Matías Gulf, Argentina) is presented as a case study. Here, the erosion of coastal landforms eliminated archaeological sites of Middle Holocene age and the unequal development of aeolian deposits affected its temporal representation at the regional scale. Both phenomena are linked to coastal dynamics and relative changes in sea level. Other environmental factors are also shown to reduce the evidence available from rockshelters in the area for chronology building. Thus, geomorphological dynamics expressed in different types of discontinuities have produced biases that affect the chronologies obtained for the human occupation of this region.

Keywords Coastal geomorphology · Biases in human occupation chronologies · Northern patagonia · Middle–late holocene

Introduction

Knowledge of the geomorphological dynamics of a landscape of human occupation is necessary for understanding of the radiocarbon chronologies obtained from archaeological materials. This is because geomorphological processes control the preservation of the landforms and stratigraphic units that contain cultural evidence (Gladfelter 1977; Waters and Kuehn 1996; Bettis and Mandel 2002; Holdaway et al.

C. M. Favier Dubois (✉)
Facultad de Ciencias Sociales, Departamento de Arqueología, INCUAPA – CONICET, Universidad Nacional del Centro de la provincia de Buenos Aires (UNICEN), Av. Del Valle 5737, B7400JWI Olavarría, Argentina
e-mail: cfavier3@gmail.com

H. Inda Ferrero and F. García Rodríguez (eds.), *Advances in Coastal Geoarchaeology in Latin America*, The Latin American Studies Book Series,
https://doi.org/10.1007/978-3-030-17828-4_1

2005; Fanning et al. 2009). The erosion of coastal landforms used by human groups, as well as the creation of vertical discontinuities in the stratigraphy, generates temporal hiatuses, sometimes to an important degree. This highlights the need to evaluate the processes responsible for the presence or absence of archaeological records in each possible coastal scenario and to explore what deposits (or surfaces) may have preserved such records.

The main units that preserve the Patagonian coastal archaeological record are eolian sediments (i.e., dunes and sand sheets). Thus, knowing the periods they represent is key to assessment of the possible chronologies and biases inherent to the record they present. This question also applies to colluvial, fluvial, or rockshelter deposits, among others, which are part of the variability of coastal geomorphological contexts. If the stratigraphic record is not continuous, neither will the evidence it preserves be continuous. In cases where erosion has prevailed along time, archaeological materials can be superimposed on ancient surfaces (e.g., marine terraces, desert pavements) mixing elements of diverse chronology (a cumulative palimpsest according to Bailey 2007). The problem is how to obtain ages in these surface assemblages since the preservation of organic material is very difficult.

The variable preservation generated by geological processes makes the datable organic materials susceptible to biases or overrepresentations, which will be expressed in the ages obtained from them in a study area. Both situations should be considered but the biases are the most problematic. A bias in the preservation of stratigraphic units and their fossil or artefactual content is called taphonomic bias, and it has a greater effect as age increases (Surovell et al. 2009). There are other biases that affect the chronologies to be obtained in an archaeological investigation, among them the research objectives, the visibility or not of the sites, the techniques of prospecting and sampling, and the choice of the material to be dated (Waters and Kuehn 1996; Waters 2000; Holdaway et al. 2005; Fanning et al. 2009; Favier Dubois 2013; Martínez et al. 2015). However, it is herein considered important to highlight in our case study the processes that affect the preservation of units that contain archaeological evidence which can be dated by ^{14}C in a coastal landscape. In this regard, some geomorphological factors that may explain the discontinuities observed in the radiocarbon chronologies obtained in the San Matías Gulf are discussed.

Study Area

The study area includes the northern and western coast of the San Matías Gulf, Río Negro province, northern Atlantic coast of Patagonia, Argentina (Fig. 1). The region has a semiarid climate, with an average temperature of 15 °C and rainfall of less than 300 mm per year. The vegetation is shrubby and corresponds to the ecological province of "Monte" (Cabrera and Willink 1980), where jarillas (*Larrea* spp.) and chañar (*Geoffroea decorticans*) are conspicuous. The tides have a semidiurnal regime and a macrotidal range, with average values of about 6 m and maxima of about 9 m, recorded in the main coastal inlet of the gulf, the San Antonio Bay (Servicio de Hidro-

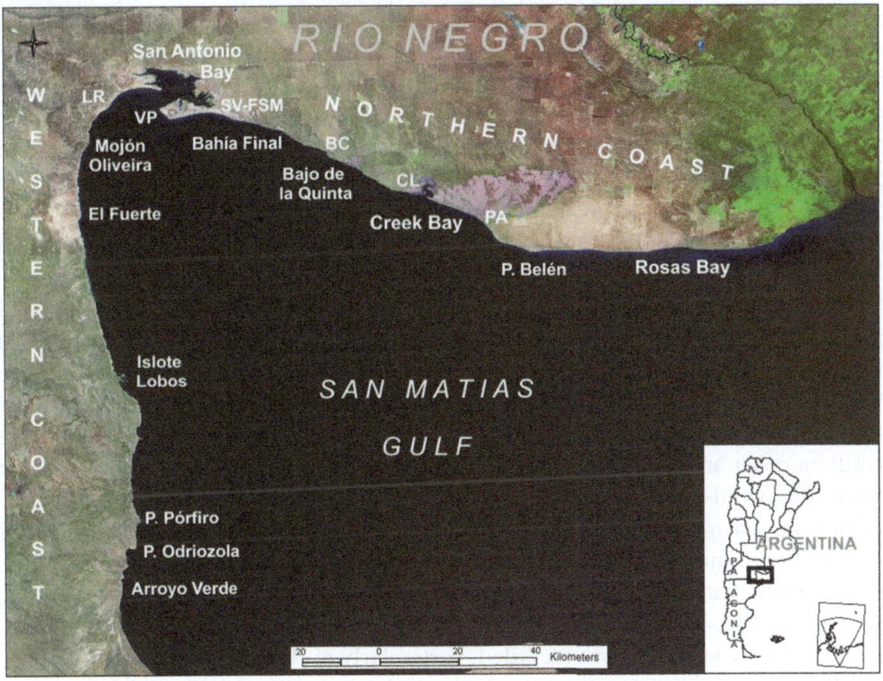

Fig. 1 Satellital image of the San Matías Gulf in the northern Atlantic Ocean coast of Patagonia (Argentina) where the localities mentioned in the text are indicated

grafía Naval 2017) (Fig. 1). The prevailing winds blow from the western quadrant (NW, W, and SW) and become the strongest between September and January, with maximum historic speeds of about 90–100 km/h recorded for the SW (Crisolitl and Pahlssa Campá 1973). These strong winds shape the direction of the dune fields that develop from the beaches to the interior of the continent, particularly in the northern coast of the gulf.

Geological and Geomorphological Background

There are noticeable geological and geomorphological differences between the northern and western coast of the San Matías Gulf. In the northern coast, an ancient raised fluvial plain is interrupted by topographic hollows (locally known as "bajos") such as the Gran Bajo del Gualicho and the Bajo de San Antonio, among others smaller ones (González Díaz and Malagnino 1984; Kokot and Favier Dubois 2017). The contact of this plain with the sea has generated a predominantly cliffed coast, with outcrops of Tertiary sandstones and siltstones of the Patagonia and Río Negro formations (Gelós et al. 1988). These rocks are friable and allow a rapid retreat of the cliff face due to the action of waves. The coastal entrances of this northern littoral develop mainly in relation to the aforementioned "bajos", as in the case of Caleta de los Loros (CL), Creek Bay, Bajo de la Quinta and San Antonio Bay (Fig. 1). At bays and other coastal inlets, dune fields develop that migrate in a NE direction; the main one is a huge sandy tongue next to Creek bay, which has reached 45 km inland in a SW-NE direction (Fig. 1). There are also other important eolian deposits in Bajo de la Quinta and Rosas Bay, as well as smaller patches of dunes between Villarino Peninsula (VP) and Bajo de la Quinta (Fig. 1). Most of them contain abundant archaeological evidence, principally shell middens.

At the western coast of the San Matías Gulf, on the other hand, the Somún Curá basaltic plateau is the prominent feature comprising a structural lava plateau that descends gradually towards the sea in the shape of pediments (González Díaz and Malagnino 1984). In these littoral zones, small estuaries are developed with sandy shores and gravel beach ridges. Eolian landforms are not as abundant as on the northern coast and are mainly located on coastal points projecting to the east that receive the winds that carry sand from the beaches located SW. This occurs in Punta Odriozola, Arroyo Verde, and other smaller localities (Fig. 1). In the southern sector of this littoral, there are hilly areas where rocks, mostly volcanic, folded, and faulty, generate a high and irregular coast. These rocks are mainly rhyolites of the Jurassic Marifil volcanic complex (Busteros et al. 1998). There, the formation of small rock cavities and rockshelters was recorded, particularly at Punta Pórfido (Fig. 1).

The archaeological localities recognized along the San Matías Gulf are found mainly in dunes and sand sheets overlying marine terraces. The Pleistocene marine terraces have heights greater than 10 m above the current sea level whereas the Holocene ones are always below this level (Schellmann and Radtke 2010; Favier Dubois and Kokot 2011; Kokot and Chomnalez 2012; Kokot and Favier Dubois 2017). These terraces show changes in the relative sea level that have affected the whole region. Radiocarbon ages obtained on biogenic deposits along the Atlantic Patagonian coast indicate that the Holocene maximum marine level had been reached by about 7000 ^{14}C yr BP (Codignotto et al. 1987; Rutter et al. 1989; Codignotto et al. 1992; Rostami et al. 2000; Schellmann and Radtke 2010; Bini et al. 2017).

Subsequent to this transgressive maximum, a marine decline is recorded in several phases. Codignotto et al. (1990) detected a period of relative stability of sea level between 6500 and 3800 ^{14}C yr BP, in Bahía Solano (province of Chubut), and then a decrease of 1.5 m between such date and 2700 ^{14}C yr BP. Monti (2000) pointed out this decrease in Bahía Engaño (province of Chubut) after 3200 ^{14}C yr BP. Subsequent work carried out along the Patagonian Atlantic coast by Schellmann and Radtke (2010) in different marine entrances located between northern Chubut province and the southern part of Santa Cruz province obtained evidence of two important sea level falls. The first occurred around 6000 yr cal BP, followed by a period of stability, and then a second sea level fall, more marked, took place around 2600 cal yr BP (Schellmann and Radtke 2010). These two pulses of decline of the relative sea level favored coastal accretion processes and the development of dunes along the San Matías Gulf.

Materials and Methods

Geomorphological units of the San Matías Gulf that contain archaeological materials have been examined. With this purpose, a bibliographic compilation of the geological and geomorphological background of the area was carried out, as well as the analysis of satellite images (Google Earth 2016) and aerial photographs (IFTA for Shell, 1959, scale 1:20,000). Field observations were also made, which were added to the information obtained by previous geoarchaeological studies (Favier Dubois and Kokot 2011; Favier Dubois and Scartascini 2012; Kokot and Favier Dubois 2017). The recognized geomorphological units are the following: (1) littoral sandy dunes (of the parabolic and barjanoid types) and sandy sheets; (2) colluvial and colluvio-eolian deposits from the cliff slopes; (3) marine and fluvial terrace surfaces; (4) a rockshelter. In these units were obtained numerous ^{14}C dates for organic materials of archaeological origin: shells (Mytilidae), charcoal, bones of marine or terrestrial fauna, and white croacker (*Micropogonias furnieri*) otoliths. The radiocarbon dates for human remains were not considered for this study because these remains are intrusive in the deposits that contain them. Thus, 69 dates obtained at twelve archaeological localities on the northern coast (i.e., San Antonio Bay, Villarino Peninsula—VP-, Saco Viejo—SV-, Faro San Matías—FSM-, Bahía Final, Barranca de los Concheros—BC-, Bajo de la Quinta, Caleta de los Loros—CL-, Creek Bay, Paesani—PA-, Promontorio Belén and Rosas Bay) and 20 dates from six archaeological localities on the western coast of the San Matías Gulf (i.e., La Rinconada—LR-, Mojón Oliveira, Islote Lobos, Punta Pórfido, Punta Odriozola and Arroyo Verde) were included (Fig. 1). The bulk of these radiocarbon dates has been published in previous papers, and only a few are added in this contribution. All these data were entered into the Cal Pal program (2007 version), calibrated to one sigma and graphed (Figs. 4 and 6). In the case of the Mytilidae valves (*Mytilus platensis* and *Aulacomya ater*), a reservoir effect value of 266 years, locally obtained, was used for correction (Favier Dubois and Jull 2017). In the case of other marine species (remains of fish and sea lions), the value of 400 years ($\Delta R = 0$) corresponding to the global marine average was used. The Holocene was subdivided taking into account the limits established by the International Union of Geological Sciences (IUGS) in 2018 that differentiate the Greenlandian (11,700 yr b2 k), Northgrippian (8326 yr b2 k), and Meghalayan (4200 yr before 1950) stages (http://www.iugs.org/) that correspond to the Early, Middle, and Late Holocene.

Results

Northern Coast of the San Matías Gulf

This coast is the most studied in the San Matías Gulf and has the largest number of archaeological sites and ^{14}C dates (Table 1). The ^{14}C ages obtained a range from *ca.*

Table 1 Radiocarbon dates obtained in the northern coast of the San Matías Gulf

Geomorphological context	Dated material	Number of dates	Rounded ^{14}C age ranges (yr BP)	References
Dunes and sand sheets	Charcoal	17	5200–450	Favier Dubois et al. (2009) Borella and Cruz (2012) Favier Dubois (2013)
	Shells	22	5700–920	
	Bone/egg	5	4600–435	
Colluvio-eolian talus	Charcoal	5	3400–1700	Favier Dubois et al. (2009) Favier Dubois and Jull (2017)
	Shells	5	3600–2100	
Marine/fluvial terrace surfaces	Fish otoliths	15	6300–1100	Scartascini et al. (2009) Favier Dubois and Scartascini (2012)

6300 to 450 yr BP. Here, the contexts of preservation of archaeological materials are the following:

(a) Dunes and sand sheets: in these deposits, diverse organic materials have been dated coming from shell middens, which are abundant along this entire littoral zone.
(b) Colluvio-eolian talus: these are the slopes of marine paleo-cliffs in which anthropic lenses of valves and charcoal have been preserved. The location that best represents this context is Barranca de los Concheros (BC) (Fig. 2).
(c) Marine/fluvial terrace surfaces: these are surface contexts where organic evidence is usually not preserved except for very resistant materials. This is the case of the white croacker otoliths, high density mineralized organic structures which have sizes from 1 to 3 cm. They were found mainly in terraces from San Antonio Bay, Bajo de la Quinta, and Creek Bay.

Western Coast of the San Matías Gulf

The western coast has ^{14}C dates ranging from 6600 to 715 yr BP (Table 2). The contexts of preservation of archaeological materials in this littoral are the following:

- Dunes and sand sheets: the materials dated in these deposits come from shell middens located along this litoral but concentrated at Islote Lobos, Punta Odriozola, and Arroyo Verde (Fig. 3).
- Colluvial talus: this develops at the foot of the hill ranges at Punta Pórfido, where the surface runoff has excavated erosion gullies which runs towards the sea. In the

Fig. 2 Colluvio-eolian talus with interstratified shell midden lenses at Barranca de los Concheros locality, northern coast of the San Matías Gulf

Table 2 Radiocarbon dates obtained in the western coast of the San Matías Gulf

Geomorphological context	Dated material	Number of dating	Rounded ^{14}C age ranges or dates (yr BP)	References
Dunes and sand sheets	Charcoal	7	2400–715	Favier Dubois (2013) Borella et al. (2015) Borella et al. (2018) This paper
	Shells	9	8 dates between 3600 and 920 2210 ± 80 (LP3366)	
Colluvial talus	Shells	1	3680 ± 90 (LP3529)	This paper
Rockshelter (Alero 2)	Charcoal	3	2 dates of 6600 and 4500 4800 ± 100 (LP3569)	Carranza and Cardillo (2018) This paper

Fig. 3 Geomorphological context of the Arroyo Verde 1 site located in a deflated dune in the western coast of the San Matías Gulf (Drone image: 36 m high)

exposed profiles, anthropic lenses of valves can be seen. At the moment, only one of these lenses has been dated.

- Rockshelter: a rockshelter (Alero 2) has been excavated in Punta Pórfido with a sequence rising 1 m above the basal rocks. The stratigraphic sequence shows a succession of numerous charcoal lenses, three of which were dated, coming from the top, middle, and the bottom of the sequence.

Discussion

The Eolian Deposits

The development of eolian deposits was conditioned by coastal geomorphological factors, which in the case of the San Matías Gulf were linked to sea level oscillations. From the transgressive maximum (ca. 7000 yr BP) until the second pulse of sea level fall (ca. 3200/2600 yr BP), there was a moderate development of mainly parabolic dunes on the northern coast of the San Matías Gulf. This occurred in coastal inlets such as San Antonio Bay, Bajo de la Quinta, Creek Bay, and Rosas Bay, where

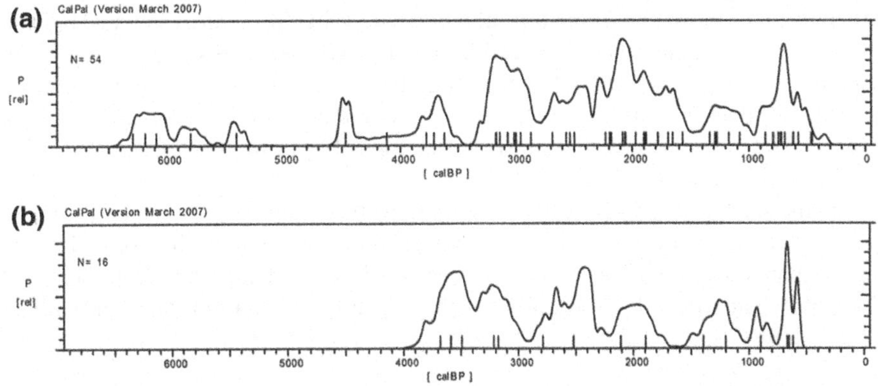

Fig. 4 Chronologies obtained in eolian and colluvial deposits of the San Matías Gulf. **a** At the northern coast (ca. 6300–450 cal yr BP) the representation of the Middle Holocene is scarce, with only 6 dates as opposed to 48 for the Late Holocene. **b** At the western coast (ca. 3700–600 cal yr BP) no Middle Holocene ages were obtained in these deposits

beach ridges were accumulated making sand available for the development of dunes during this period. The coastal sectors outside of these marine inlets, which account for approximately 70% of this coastline, suffered the predominant effect of erosion during those times, with the great retreat of the cliffs and the erosion of many archaeological sites. This explains why only one Middle Holocene shell midden (from ca. 5600 yr BP) was detected outside of the aforementioned inlets on the northern coast of the San Matías Gulf (Favier Dubois 2013). Following the second pulse of sea level fall (ca. 3200/2600 yr BP), coastal accretion increased remarkably along the coast of the San Matías Gulf and the development of beach ridges provided a great amount of sand. This favored the formation of large bodies of barjanoid dunes that spread into many coastal areas. The proliferation of these deposits favored the preservation of the archaeological evidence of the Late Holocene (Table 1, Fig. 4a) that is overrepresented with respect to the previous period. The great development of wind-blown deposits during this late stage also affected the visibility of the older ones, which are less obtrusive.

On the western coast of the San Matías Gulf, as on the northern coast, after the second pulse of sea level fall, the beach ridges supply of sand increased. This favored also the formation of Late Holocene dunes and a greater preservation of shell middens corresponding to this time here (Table 2, Figs. 3 and 4b).

The Colluvial Talus

These deposits consist of the slopes of marine paleo-cliffs that developed during the sea regression. After the first sea level fall (ca. 6000 yr BP), but mainly after the second (ca. 3200/2600 yr BP), coastal accretion separated many cliffs from the effects of marine erosion. From this moment, a slope of colluvium developed in front of the cliffs, in some cases increased by wind-blown sediments (Fig. 2). It can be seen in Table 1 that the dates obtained for archaeological materials in these contexts are all later than that of the second pulse of marine descent (ca. 3200/2600 yr BP). This is related to the fact that most of the slopes began to evolve after that more marked pulse of sea level fall.

On the western coast, the hill slopes might well contain very old materials. However, the archaeological material was recovered in erosion gullies that formed on the slopes that descend directly towards the sea and developed after the second sea level fall pulse. This links them to this final regressive phase and to a later chronology (Table 2, Fig. 4b).

Marine/Fluvial Terrace Surfaces

In this case, organic evidence has been preserved on the surface despite the lack of a sedimentary matrix. The evidence consists of otoliths, associated with stone net weights that have survived in large quantities on marine and fluvial terraces, which constituted intensive net fishing scenarios during the Middle Holocene (Favier Dubois and Kokot 2011; Favier Dubois and Scartascini 2012). These ancient scenarios have been preserved in northern coastal inlets such as San Antonio Bay, Bajo de la Quinta, and Creek Bay. Otoliths are structures resistant to the absence of sedimentary matrix and are the main source of chronological data for the Middle Holocene in the whole San Matías Gulf (Fig. 6a). Without these structures, there could be a large bias in the human signal during that time.

Rockshelter Deposits

Rockshelters have not been eroded by the sea transgression and are good sedimentary traps. They do not guarantee continuity in sedimentation but may represent different time periods including early stages of settlement. However, for the sequence of deposits of the excavated rockshelter, Alero 2, all dates (which includes the top and the base of the sequence) correspond to the Middle Holocene, indicating that the preserved fill corresponds only to that time (Fig. 6b). It is interesting to note that such a chronology has not yet been identified for the eolian or colluvial deposits of this littoral (Fig. 4b), some of them very close to Alero 2. It is also remarkable that this rockshelter has such a temporally limited sequence of fills; in relation to this we have observed that the sedimentary fill has been removed from some rockshelters of this area in recent times. This situation is mentioned little in the literature but shows that the sediment accumulation can be reversible, with not only filling cycles but also sediment washing processes that could eliminate evidence of previous occupations. Also, as the fills rise, the sediments may not be able to remain in the rockshelter but

Fig. 5 View of the Alero 2 and its geomorphological context. The rockshelter front is about 10 m long

are rather forced to creep outwards. This seems to be the case in Alero 2 (Fig. 5), at whose surface has an outward inclination of 16°–18°, values that favor the downwards

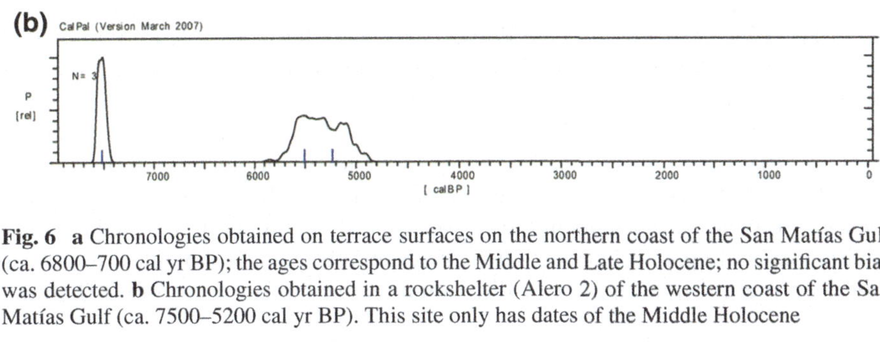

Fig. 6 **a** Chronologies obtained on terrace surfaces on the northern coast of the San Matías Gulf (ca. 6800–700 cal yr BP); the ages correspond to the Middle and Late Holocene; no significant bias was detected. **b** Chronologies obtained in a rockshelter (Alero 2) of the western coast of the San Matías Gulf (ca. 7500–5200 cal yr BP). This site only has dates of the Middle Holocene

creep of deposits. These and other processes still to be addressed may contribute to explain the virtual absence of Late Holocene sediments at least in a sector of the Alero 2. In sum, a rockshelter suffers geomorphological biases different from those of other sedimentary environments that contain cultural evidence, but they are present nevertheless and they are sometimes significant (see Farrand 2001).

Conclusions

Throughout the Holocene, the evolution of the coastal landscape generated numerous landforms suitable for use by the North Patagonian hunter-gatherers. These spaces experienced various geomorphological dynamics that each had their own impact on the preservation of the archaeological record; coastal spaces could appear and disappear, and variable erosion–sedimentation pulses occurred over time. The age diagrams show that those chronologies difficult to obtain in eolian and colluvial deposits (Fig. 4a, b) have been provided by resistant materials on terraces and rockshelter sequences in the San Matías Gulf (Fig. 6a, b). This demonstrates that the study of diverse geomorphological contexts and the dating of varied organic materials helps to offer a more complete vision (without considering other biases) of the chronology of use of coastal spaces. Since there exists a disparity in preservation, it is necessary to focus on the carrier units rather than on the cultural materials themselves in order to evaluate and contextualize better the temporal tendencies in the use of space. At the San Matías Gulf, geomorphological factors are responsible for the elimination of coastal sites, for the relative abundance of wind deposits for each occupation period, and for the erosion processes that affected rockshelter sequences. These geomorphological dynamics are not a peculiarity of the coast of the San Matías Gulf, but a natural and predictable consequence of how environmental factors act in many littoral (and other) spaces. This fact significantly affects our vision of the history of human occupation from the first peopling of a region. In relation to this, it is known that there is a significant geomorphological factor contributing to the difficulty in locating early coastal occupations since these were largely eliminated by the Middle Holocene marine transgression (e.g., Bailey and Milner 2002). In the case of the western coast of the San Matías Gulf, the presence of a 11,000 yr cal BP coastline close to the present shore (Ponce et al. 2011) would provide evidence of early occupations at places not affected by the marine transgression. However, in the excavated rockshelter at Punta Pórfido, it has not been possible to overcome at the moment the "Middle Holocene barrier" that show radiocarbon dates at the whole Patagonian Atlantic Ocean coast (Zangrando 2018).

The preserved (and detected) sedimentation episodes containing cultural evidence are those that are going to finally be expressed in the occupation chronology of those landscapes. The studies are then conditioned (with few exceptions) by the presence of these carrier units of the archaeological record in different sedimentary environments. Temporal hiatuses often correspond to geomorphological processes modeling preservation and not to discontinuities in human occupations (Farrand

1993; Waters and Kuehn 1996; Bettis and Mandel 2002; Holdaway et al. 2005; Fanning et al. 2009; Favier Dubois et al. 2017), and therefore the study of these processes provides a necessary framework to evaluate occupation chronologies in any region under consideration.

Acknowledgements I thank Professor Hugo Inda Ferrero and Professor Felipe García-Rodríguez (editors) for its important role in the publication of this special volume. I also thank Richard Lunnis and Daniel Rafuse for the English revision of the manuscript. This research was funded by grants from the CONICET (National Research Council of Argentina, PIP 00756 and 00589 projects) and ANPCyT (National Agency for Science and Technology of Argentina, PICT 2013 1128).

References

Bailey G (2007) Time perspectives, palimpsests and the archaeology of time. J Anthropol Archaeol 26:198–223

Bailey G, Milner N (2002) Coastal hunter-gatherers and social evolution: marginal or central? Before Farm 2002/3 4(1):129–150

Bettis EA III, Mandel RD (2002) The effects of temporal and spatial patterns of Holocene erosion and alluviation on the archaeological record of the Central and Eastern Great Plains, USA. Geoarchaeology 17(2):141–154

Bini M, Isola I, Zanchetta G, Pappalardo M, Ribolini A, Ragaini L, Baroni C, Boretto G, Fuck E, Morigi C, Salvatore MC, Bassi D, Marzaioli F, Terrasi F (2017) Mid-Holocene relative sea-level changes along Atlantic Patagonia: new data from Camarones, Chubut, Argentina. Holocene 28(1):56–64

Borella F, Cruz I (2012) Taphonomic evaluation of penguin (Spheniscidae) remains at a shell-midden on the northern coast of Patagonia (San Matías Gulf, Río Negro, Argentina). Quatern Int 278:45–50

Borella F, Cardillo M, Favier Dubois CM, Alberti J (2015) Nuevas investigaciones arqueológicas entre Punta Pórfido y Punta Odriozola: implicancias para el entendimiento de la dinámica de las ocupaciones humanas en la costa oeste del golfo San Matías (Río Negro). Relaciones de la Sociedad Argentina de Antropología 40(1):233–252. Buenos Aires

Borella F, Cardillo M, Scartascini F, Alberti J, Steffan PG, Favier Dubois CM (unpublished) Primeros resultados de las investigaciones arqueológicas en el área del Complejo Islote Lobos, costa oeste del golfo San Matías (Río Negro) (ms under revision, 2018)

Busteros A, Giacosa R, Lema H (1998) Hoja Geológica 4166-IV Sierra Grande, Provincia de Río Negro. Servicio Geológico Minero Argentino, Buenos Aires, p 75

Cabrera AL, Willink A (1980) Biogeografía de América Latina. OEA monographs. OEA, Washington D.C.

Carranza E, Cardillo M (2018) Desprendimientos rocosos y ruido tafonómico: trabajo experimental en el Alero 2 de Punta Pórfido, Golfo San Matías (Provincia de Río Negro, Argentina). Revista Chúngara (in press)

Codignotto JO, Beros CA, Trebino LG (1987) Nuevo método cronoestratigráfico, morfocronología en secuencia deposicional cordoniforme. Revista de la Asociación Geológica Argentina 42(3–4):462–468. Buenos Aires

Codignotto JO, Cesari O, Beros CA (1990) Morfocronología secuencial evolutiva holocena en bahía Solano, Chubut. Revista de la Asociación Geológica Argentina 45(3–4):205–212. Buenos Aires

Codignotto JO, Kokot RR, Marcomini SC (1992) Neotectonism and sea level changes in the coastal zone of Argentina. J Coastal Res 8:125–133

Crisoliti LE, Pahissa Campá ME (1973) Climatología. In: Olivier SR (ed) Relevamiento Ecológico y tipificación de las comunidades del Litoral Marítimo de la Provincia de Río Negro, con especial

referencia al establecimiento de áreas de cultivo para especies de interés comercial. Instituto de Biología Marina. Asesoría de desarrollo de Río Negro. Consejo Federal de Inversiones, unpublished report, pp 11–88. Buenos Aires
Fanning P, Holdaway SJ, Rhodes EJ, Bryant TG (2009) The surface archaeological record in arid Australia: Geomorphic controls on preservation, exposure, and visibility. Geoarchaeol Int J 24(2):121–146
Farrand WR (1993) Discontinuity in the stratigraphic record: snapshots from Franchthi cave. In: Goldberg P, Nash DT, Petraglia MD (eds) Formation processes in archaeological context, monographs in world archaeology, vol 17, pp 85–96. Prehistory Press, Madison, Wisconsin
Farrand WR (2001) Archaeological sediments in rockshelters and caves. In: Stein JK, Farrand WR (eds) Sediments in archaeological context. University of Utah Press, 218 pp
Favier Dubois CM (2013) Hacia una cronología del uso del espacio en la costa norte del golfo San Matías (Río Negro, Argentina): sesgos geológicos e indicadores temporales. In: Zangrando F Barberena R, Gil A (eds) Tendencias teórico-metodológicas y casos de estudio en la arqueología de Patagonia, pp 87–96. Editorial Altuna, Buenos Aires
Favier Dubois CM, Kokot RR (2011) Changing scenarios in Bajo de la Quinta (San Matías Gulf, Northern Patagonia, Argentina): impact of geomorphologic processes in subsistence and human use of coastal habitats. Quatern Int 245(1):103–110
Favier Dubois CM, Scartascini F (2012) Intensive fishery scenarios on the North Patagonian coast (Río Negro, Argentina) during the Mid-Holocene. Quatern Int 256(1):62–70
Favier Dubois CM, Jull T (2017) Marine reservoir effect values from shell-middens of San Matías gulf, Northern Patagonia (Argentina): a 5000-yr record. Radiocarbon 60(2):1–6
Favier Dubois CM, Borella F, Tykot RH (2009) Explorando tendencias temporales en el uso del espacio y los recursos marinos en el Golfo San Matías (Río Negro). In: Santiago F, Salemme M, Alvarez M, Piana E, Vázquez M, Mansur ME (eds) Arqueología de Patagonia: una mirada desde el último confín, tomo II, pp 985–997. Editorial Utopías, Ushuaia, Tierra del Fuego
Favier Dubois CM, Massigoge A, Messineo P (2017) El Holoceno medio en valles del Area Interserrana Bonaerense, ¿escasez de sitios o de unidades portadoras? una perspectiva geoarqueológica. Revista del Museo de Antropología 10(2):19–34. Buenos Aires
Gelós EM, Spagnuolo JO, Schillizzi RA (1988) Las unidades morfológicas de la costa norte de Golfo San Matías y su evolución. Revista de la Asociación Geológica Argentina 43(3):315–326. Buenos Aires
Gladfelter BG (1977) Geoarchaeology: the geomorphologist and archaeology. Amer Antiq 42:519–538
González Díaz E, Malagnino E (1984) Geomorfología de la provincia de Río Negro. Actas del IX Congreso Geológico Argentino. Secretaría de Minería, Buenos Aires, pp 1–159
Holdaway SJ, Fanning PC, Shiner J (2005) Absence of evidence or evidence of absence? Understanding the chronology of indigenous occupation of western New South Wales, Australia. Archaeol Ocean 40:33–49
Kokot R, Chomnalez F (2012) Retroceso de la línea de costa en Las Grutas, provincia de Río Negro. Revista de la Asociación Geológica Argentina 69(1):88–96. Buenos Aires
Kokot R, Favier Dubois CM (2017) Evolución geomorfológica de la bahía de San Antonio, Provincia de Río Negro. Revista de la Asociación Geológica Argentina 74(3):315–325. Buenos Aires
Martínez G, Prates L, Flensborg G, Stoessel G, Alcaráz AP, Bayala P (2015) Radiocarbon trends in the Pampean region (Argentina). Biases and demographic patterns during the final Late Pleistocene and Holocene. Quatern Int 356:89–110
Monti AJ (2000) Edades ^{14}C y ciclicidad de la acreción en depósitos costeros elevados, Bahía Engaño, Chubut. Revista de la Asociación Geológica Argentina 55(4):403–406. Buenos Aires
Ponce JF, Rabassa J, Coronato A, Borromei AM (2011) Palaeogeographical evolution of the Atlantic coast of Pampa and Patagonia from the last glacial maximum to the Middle Holocene. Biol J Lin Soc 103:363–379

Rostami K, Peltier WR, Mangini A (2000) Quaternary marine terraces, sea level changes and uplift history of Patagonia, Argentina: comparison with predictions of the ICE-4G (VM2) model of the global process of glacial isostatic adjustment. Quatern Sci Rev 19(4):1495–1525

Rutter N, Schnack EJ, del Rio J, Fasano JL, Isla FI, Radtke U (1989) Correlation and dating of quaternary littoral zones along the Patagonian coast, Argentina. Quatern Sci Rev 8:213–234

Servicio de Hidrografía Naval (2017) Tabla de Mareas. Servicio de Hidrografía Naval, Armada Argentina. (http://www.hidro.gov.ar/oceanografia/Tmareas/Form_Tmareas.asp). Buenos Aires

Scartascini FL, Charo M, Volpedo A (2009) Caracterización de las estrategias de obtención de recursos ícticos a partir del análisis de otolitos. El caso de la costa norte del Golfo San Matías. In: Santiago F, Salemme M, Alvarez M, Piana E, Vázquez M, Mansur ME (eds) Arqueología de Patagonia: una mirada desde el último confín, tomo II, pp 845–852. Editorial Utopías, Ushuaia, Tierra del Fuego

Schellmann G, Radtke U (2010) Timing and magnitude of Holocene sea-level changes along the middle and south Patagonian Atlantic coast derived from beach ridge systems, litoral terraces and valley-mouth terraces. Earth Sci Rev 103:1–30

Surovell T, Finley JB, Smith GM, Brantingham J, Kelly R (2009) Correcting temporal frequency distributions for taphonomic bias. J Archaeol Sci 36:1715–1724

Waters MR, 2000. Alluvial stratigraphy and geoarchaeology in the American Southwest. Geoarchaeol Int J 15(6):537–557

Waters MR, Kuehn DD (1996) The geoarchaeology of place: the effect of geological processes on the preservation and interpretation of the archaeological record. Am Antiq 61(3):483–497

Zangrando AF (2018) Poblamiento temprano y arqueología de costas en Patagonia y Tierra del Fuego: vacío de información, preconceptos y perspectivas. Intersecciones en Antropología (in press). Buenos Aires

A Basin-Wide Assessment of Natural Dynamics and Modern Human Impacts on the Visibility and Conservation of Coastal Archaeological Sites in the Atlantic Coast of Uruguay

Hugo Inda Ferrero and Laura del Puerto

Abstract Coastal geoarchaeology has the potential to transform itself in a leading discipline with the capability to provide useful insights for the adaptation of coastal populations to rising seas and changing environments in a future that is getting alarmingly closer. In this contribution, we present the results of a research that combined the state of the art for Uruguayan coastal archaeology with Holocene paleoenvironmental evolution and modern anthropogenic impact from a basin-wide perspective. Our findings identified a direct relationship between sea level evolution and climate dynamics with the scarcity of human evidences for the Greenlandian and Northgrippian stages in coastal environments, where harsher environmental conditions could have discouraged the exploitation of these landscapes but also erased these earlier evidences after sites abandonment. The prolific archaeological record for the Meghalayan was explained by the establishment of more favorable climatic conditions and relatively stable environments promoting the spread and settlement across coastal landscapes. Despite this fact, modern anthropogenic impact poses a growing threat to this cultural heritage, as market-driven real estate development over capes and dune fields and cropland expansion over grasslands are occupying critical areas. In this scenario, this work identified less impacted areas where exhaustive archaeological surveys coupled with other conservation measures are urgently needed.

Keywords Coastal geoarchaeology · Anthropogenic impact · Natural variability · Holocene · Uruguay

---

H. Inda Ferrero (✉) · L. del Puerto
Centro Universitario Regional del Este (CURE), Universidad de la República, Ruta 9 y Ruta 15, 27000 Rocha, Uruguay
e-mail: hinda@cure.edu.uy

H. Inda Ferrero and F. García Rodríguez (eds.), *Advances in Coastal Geoarchaeology in Latin America*, The Latin American Studies Book Series,
https://doi.org/10.1007/978-3-030-17828-4_2

Introduction

Coastal environments are highly dynamic ecosystems both in time and space; their transitional nature between continental and marine realms gives place to unparalleled species richness and biodiversity (Kellog and Mohriak 2001; Watanabe et al. 2018). These ecologically highly productive environments attracted human populations to settle in the littoral areas around the globe since the dawn of mankind as it is witnessed by several ancient coastal sites (Inda et al. 2011; Werz et al. 2014). Thus, prehistoric coastal settlements could hold the key to understand how humans interacted with changing environments in the past. Such adaptations and feedbacks are extremely relevant in order to unveil ancient successful practices developed by humans to survive and prosper in coastal landscapes. Furthermore, current and future mitigation and adaptation measures to sea level rise and other climate change effects in coastal areas could gain insights from these ancient developments allowing, at the same time, to generate early warning signals for non-sustainable actions (Bradje and Erlandson 2013; Inda et al. 2016).

In this context, several lines of evidence must be conjugated in order to shed light over persistent key questions: How natural dynamics shaped coastal environment? How past human societies faced those changing scenarios? What evidence did survive considering site formation processes and modern human impact over such areas? The Atlantic coast of Uruguay is a privileged scenario to fill these gaps of knowledge, because it holds some high-resolution records of Holocene local environmental responses to global changes and a prolific archaeological record of human adaptative mechanisms. Moreover, the area also has a gradient of modern anthropogenic disturbance that ranges from severely modified environments to almost pristine beaches, what allows not only to approach modern impact over archaeological sites preservation and visibility but also to propose future actions for less impacted areas.

In order to disentangle the complex feedback mechanisms between humans and coastal environments across the Holocene, this contribution will first introduce a review of the state of the art of Uruguayan coastal Archaeology. Second, a set of analyses will be developed, aimed to show how the whole picture is far more complex, because these mechanisms are permeated by modern anthropogenic modifications of such areas, affecting preservation and visibility of archaeological sites (Giannoti et al. 2017). In most cases, the scarcity of archaeological remains (or the complete lack of) in coastal scenarios should not be inferred as an evidence for the absence of human settlements, mainly because natural processes such as coastal erosion and ecological succession mechanisms could hide or even erase ancient human imprints. In addition, current anthropogenic processes such as urban coastal development must be seeing as powerful disturbing agents by their own, but also because they trigger or promote natural hazards, which could lead to unintended and unforeseen consequences in terms of archaeological sites preservation and visibility.

Holocene Coastal Dynamics and the Archaeological Record

Uruguayan archaeology has dealt with subsistence strategies and settlement patterns of prehistoric inhabitants and some contributions which also addressed the role of natural environmental evolution (Bracco et al. 2005; Inda et al. 2006a, 2011; Iriarte 2006; Iriarte et al. 2017; Milheira and Gianotti García 2018). Nevertheless, little attention was paid to understand the natural and anthropogenic processes that began once sites were abandoned, particularly regarding the incidence of current human impact over coastal landscapes (Gianotti et al. 2017).

The Atlantic coast of Uruguay did not experience submergence or uplift tectonic processes during the Holocene (Isla and Angulo 2016) which means that sea level change was the main driver of coastal dynamics and evolution. Regional research efforts developed in Uruguay, SE Brazil, and Argentina (Angulo et al. 2006; Bracco et al. 2014; Cavallotto et al. 2004; Isla 1989; Martin and Suguio 1992; Martínez and Rojas 2013) showed remarkable agreement about inferred sea level curves for this part of South America. These findings, coupled with intensive topographic surveys and paleolimnological multiproxy analyses through several coastal lagoon bottoms, wetlands, landforms, and exposed coastal paleosurfaces, allowed the reconstruction of past regional climate and coastal evolution and to generate the most likely coastal configurations (del Puerto et al. 2013; Inda et al. 2017) for the recently defined Holocene stages (Walker et al. 2018). Information about prehistorical inhabitants of the coast, on the other hand, has been benefited through the integration of several sources, including historical chronicles written by European explorers since the sixteenth century, collector findings since the early twentieth century and recent archaeological research (Arredondo 1958; Gianotti et al. 2017; Inda et al. 2017; Maeso 1977). Coupling inferred paleo scenarios with cultural data not only allowed us to approach adaptative mechanisms in terms of settlement patterns and subsistence strategies, but also to get insights about archaeological sites visibility, preservation, and potential risk related to current and future natural and anthropogenic processes acting along the Atlantic coast.

The Greenlandian

At the onset of the Holocene sea level was around 50 m below current sea level, with a rising trend since the Last Glacial Maximum estimated in around 0.1–0.19 cm/year (Inda et al. 2011). At this time, a vast portion of more than 70,000 km^2 of current estuarine and oceanic bottom was part of a river valley, known as the Paleo Paraná-Uruguay system, whose mouth was located at the edge of the continental shelf (Inda et al. 2017; Rosa et al. 2011).

Human footprint for this period is almost nonexistent, with only one site reported so far, located in the after-shore plains of the Atlantic coast, in the Negra Lagoon basin (López Mazz 2013) dated at 8510 $\pm$ 40 yr BP.

The Northgrippian

Ice melting from polar and continental glacier sheets coupled with rainfall and temperature increase disrupted ocean circulation and promoted that the rising sea quickly drowned this fluvial valley and, by around 7.5–7.0 ka BP, sea level reached current stands. Sea level continued the rising trend until it reached around 4.5–5 m above current sea level around 5.5 ka BP, an event referred as the Maximum Holocene Transgression (MTH hereafter; Bracco et al. 2014; Martínez and Rojas 2013; Inda et al. 2006b). During this period, the sea submerged a wide array of former *terra firme* landscapes that were sculpted during the Pleistocene (the Dolores and Libertad formations; Ubilla and Martínez 2016). This marine intrusion implied that river mouths and floodplains shifted to saline environments, as it is recorded by paleolimnological proxies from current coastal lagoons. Those lagoons functioned as shallow bays during this sea level high stand and surrounding plains as salt marshes (Bracco et al. 2005; Inda et al. 2017).

The archaeological record is, like in the previous stage, almost nonexistent for the first 2.5 millennia, with the exception of the Los Indios site, dated at 7100 ± 160 yr BP (López Mazz 2013). Near the end of this stage, the archaeological record identified a site dated at 4881 ± 55 and 4370 ± 70 ka BP in Cabo Polonio. In the high stand coast of the Castillos lagoon, just a few kilometers inland from Cabo Polonio, another archaeological site was identified, yielding an age of 4600 ± 60 (Inda et al. 2017).

The Meghalayan

The sudden climatic deterioration witnessed worldwide around 4.2 ka BP that defined the onset of the Meghalayan stage has several geomorphological and paleolimnological evidences in the Atlantic coast of Uruguay (del Puerto et al. 2006, 2013). Sea level began to fall since the MTH and this fact, coupled with a sharp decrease in rainfall promoted dramatic modifications in coastal environments. Sea level retreat exposed huge amounts of former subaquatic sand-sized sediments which, coupled with SW winds in a semiarid environment, were at the basis of current coastal configuration: the inland aeolian displacement of these sediments, favored by lesser discharges from the basins, caused a rapid enclosure of shallow bays and, simultaneously, the formation of dune fields that reached several kilometers of inland territories (Panario et al. 2014; Panario and Gutiérrez 1999; Panario and Piñeiro 1997). By around 3.0 ka BP climate shifted to a regime very similar to current climate (del Puerto et al. 2013; Garreaud et al. 2009). Such climate shift fostered the expansion of freshwater into coastal landscapes, gradually shaping current coastal configuration of arc beaches between capes and sand bars dominated by freshwater/brackish coastal lagoons.

The archaeological record for the early Meghalayan is virtually nonexistent so far. Only after sea level decrease and rainfall reached levels similar to modern times around 3.0 ka BP, with the consequent landscape transformation and gradual stabilization of the shoreline and after-shore territories, the coast began to be inhabited. Evidence for this later period is rather abundant, with sites in headlands, beach arcs, after berm wetlands, and interdune ephemeral ponds. According to available radiocarbon dates and historical chronicles, human coastal settlements continued until the arrival of European conquerors in the sixteenth century (Arredondo 1958; Gianotti et al. 2017; Inda et al. 2017).

Study Area

The study area comprises the Atlantic coast of Uruguay, from Punta del Este (the limit between the del Plata estuary and the ocean) 34° 56′ 50.89″ S, 54° 56′ 1.75″ W, and the Chuy stream mouth (the limit between Uruguay and Brazil) at 33° 46′ 00″ S 53° 24′ 00″ W. The upper limit of the coast was set by the Atlantic basin boundary as defined by the DNH (Spanish acronym for National Direction of Hydrography) covering an area of 8373 km^2 and 233 km of coastline (Fig. 1).

Soils were defined as being of Holocene age for the lower topographic areas belonging to the Angostura (dune fields and beaches) and Laguna Merin (lagoon and river coasts and floodplains) edaphic units. According to soil descriptions (Durán 1985), these units were formed after the MTH. The genesis of these soils is linked to the eustatic and geomorphologic processes described in the preceding section. The Lascano and San Luis edaphic units were developed over Pleistocene deposits, while the rest of the area has edaphic units whose genesis was related to older slope and weathering processes from predominant granitic outcrops that formed the hills and ranges of the region (Fig. 2).

Land cover information for the study area, performed for 2015, was retrieved from DINAMA's (Spanish acronym for National Direction of the Environment) web page in ESRI shapefile format (Fig. 3). Natural covers accounted for 74.05% of the total coverage: grasslands (45.94%), riparian forests (10.37%), floodplains and wetlands (8.77%), lagoons (4.46%), palm groves (1.77%), shrublands (1.37%), and dune fields and beaches (1.37%). Artificial land covers, on the other hand, were dominated by croplands (13.82%), followed by forestry plantations (10.35%), urban areas (1.51%), water reservoirs (0.22%), and quarries (0.05%).

Materials and Methods

Paleoenvironmental Data

Paleoenvironmental scenarios for the Holocene stages were reconstructed from available literature on the topic for the study area (Bracco et al. 2011; Inda et al. 2011,

Fig. 1 Study area. Background Landsat 8 satellite image mosaic

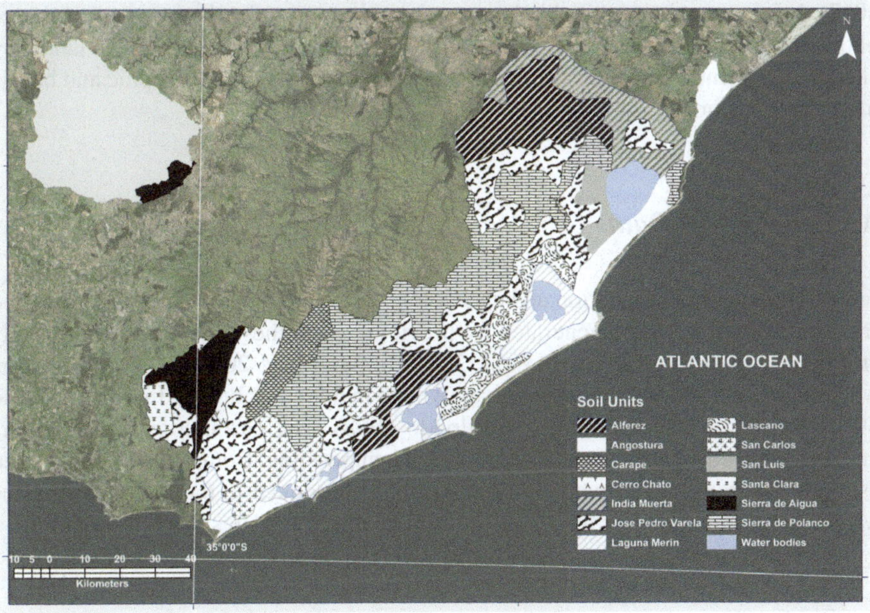

Fig. 2 Soil units of the Atlantic basin

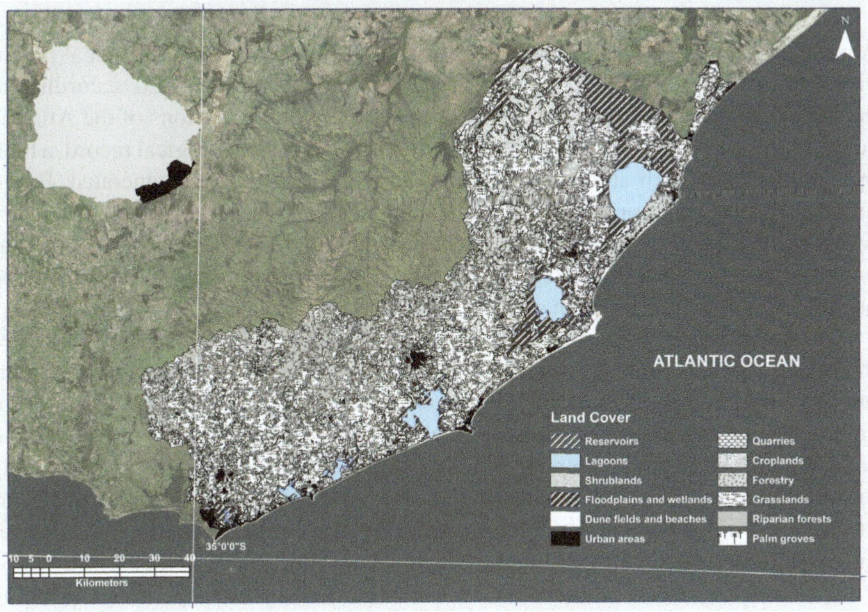

Fig. 3 Current land cover for the study area

2017). Inferences about most likely coastal configuration for these ages were based on local sea level curves (Bracco et al. 2014; Martínez and Rojas 2013), landform chronologies (Inda et al. 2017; Panario and Gutiérrez 1999), topographic and bathymetric information.

Archaeological Data

Information about the location and chronology of archaeological sites was retrieved from published literature generated by previous researches developed in the study area (Gianotti et al. 2017; Inda et al. 2017; López Mazz 2013). This information was complemented with other sources which included collector findings (Maeso 1977) and historical chronicles (Arredondo 1958). The chronology for collectors and historical chronicles sites was inferred, when possible, from artifacts considered as reliable chronological markers (pottery, first introduced in the region during the Meghalayan stage) and/or by the substratum in which the sites were allocated.

Visibility and Preservation Potential of the Archaeological Record

A multistage analysis was developed in order to approach the archaeological record visibility and preservation potential. First, the analysis was divided according to the Holocene ages and then, based on the most likely configurations of the Atlantic coast for the different ages and the information from the archaeological record, a high probability location of archaeological sites for each scenario was generated. Based on available information about the archaeological record, a 10 km inland territory was assigned for each age. This data set was converted to areas and represented as polygons in a GIS exhibiting current coastal configuration to assess visibility and preservation potential for each age.

The second stage consisted in the generation of a new information layer where the results of the first stage were intersected with the soils of the region, considering soils as a reliable and conspicuous proxy for not only past environmental and/or anthropogenic processes that explained their genesis, but also for potential future land covers that will further impact on visibility and preservation. According to available information, the soil units were divided into pre-Meghalayan and Meghalayan surfaces, as soil units described as Recent were formed after the MTH, so these younger units can only be occupied during the Meghalayan stage.

The third stage involved the addition on another information layer, corresponding to current land cover. In this case, low visibility and low preservation potential were assigned to sites located in anthropogenic land covers (croplands, forestry, urban areas) and the opposite for sites located in natural land covers.

The final stage consisted in the intersection of all the abovementioned variables, which allowed us to generate a basin-wide map of visibility and preservation potential were all requisites were merged: most probably sites allocation in terms of distance to the coastline, appropriate soil units in terms of chronology and current natural land covers.

Results

Holocene Coastal Evolution and Human Occupations

At the transition between the Pleistocene to the Holocene—the onset of the Greenlandian stage—the sea that continued a rising trend since the end of the Last Glacial Maximum around 20 ka BP, reached a coastline placed at 50 m below current sea level. This trend continued during the entire stage and until well entered the Northgrippian stage, reaching current sea level around 7 ka BP and 5 m above current sea level by 5.5 ka BP (Bracco et al. 2011, 2014). According to this Holocene evolution and available chronologies for sites in the Atlantic basin (Table 1), the most probable areas for coastal human occupations corresponding to the three Holocene stages, as shown in Fig. 4, were inferred.

With the exception of the Los Indios archaeological site, the rest of the Greenlandian sites were drowned by the rising sea in later ages; meanwhile, the Northgrippian occupations extended from an area that comprised from the current coastline to inland territories reaching even 20 m topographic contours at the MTH. Meghalayan occupations, in turn, covered an area that corresponded to current coastal environments, which is in agreement with a decreasing sea level since the MTH during the Northgrippian stage.

Edaphic Units and the Archaeological Record

In Fig. 5, the edaphic units for the inferred Holocene stages potential archaeological areas are presented (excluding the underwater Greenlandian area). Pre-Meghalayan soils roughly corresponded with the area drowned at the MTH, being soils developed over Pleistocene deposits. The areas for this stage closer to current coastline corresponded to granitic outcrops that formed stable landscape features, such as capes and beachheads. The archaeological record for these units has a chronology that further confirmed the inferred relationship between older and resilient soils and places and earlier human occupations.

Meghalayan soils, in turn, corresponded to the sea level retreat identified since the MTH that continued across this stage. Once current coastal configuration was established, human occupation was virtually continuous from the onset of the Meghalayan

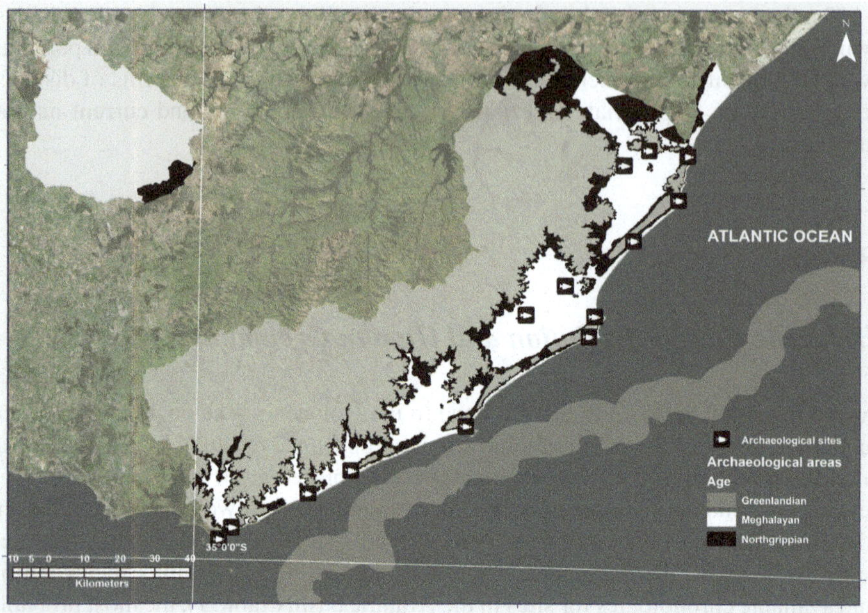

Fig. 4 Inferred archaeological areas for the Holocene stages and location on main archaeological sites. Reconstruction based on Bracco et al. (2011), Inda et al. (2011, 2017)

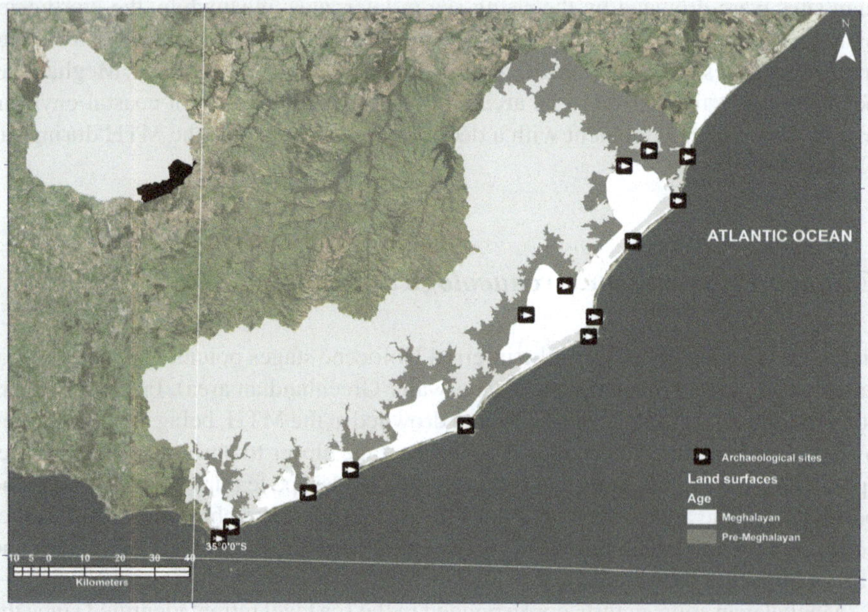

Fig. 5 Land surfaces for the pre-Meghalayan and Meghalayan stages inferred from soils of the study area and archaeological sites location

Table 1 Archeological sites of the study area, arranged using Holocene stage. Sites are presented eastward from Punta del Este to the Brazilian border

Site	Holocene stage	Chronology (years BP)
Los Indios	Greenlandian stage	8500 ± 40
Los Indios	Northgrippian stage	7100 ± 160
Cabo Polonio	Northgrippian stage	4881 ± 55 4370 ± 70
Guardia del Monte	Northgrippian stage	4600 ± 60
Punta del Este	Meghalayan stage	Inferred
La Barra	Meghalayan stage	Inferred
Manantiales	Meghalayan stage	Inferred
Laguna José Ignacio	Meghalayan stage	Inferred
Laguna de Garzón	Meghalayan stage	Inferred
Laguna de Rocha	Meghalayan stage	Inferred
Zanja Honda	Meghalayan stage	Inferred
La Paloma	Meghalayan stage	Inferred
Arachania	Meghalayan stage	Inferred
Estancia La Pedrera	Meghalayan stage	1240 ± 40
Cabo Polonio	Meghalayan stage	4100 ± 90 610 ± 65
Cráneo Marcado	Meghalayan stage	3050 ± 50
La Esmeralda	Meghalayan stage	3300 ± 40 2360 ± 70 1000 ± 70
La Moza	Meghalayan stage	2460 ± 30
Cerro Verde	Meghalayan stage	Inferred
Puntas de La Coronilla	Meghalayan stage	2930 ± 50 2740 ± 60

Source Arredondo (1958), Giannoti et al. (2017), Inda et al. (2011, 2017), López Mazz (2013), Maeso (1977)

stage to the arrival of European conquerors and settlers in the sixteenth century, as shown by radiocarbon chronologies (Table 1).

Current Land Cover and the Archaeological Record

Current land cover retrieved from the National Environmental Direction (DINAMA) country-wide survey of 2015, was divided into anthropogenic and natural covers and presented in Fig. 6 together with archaeological sites location.

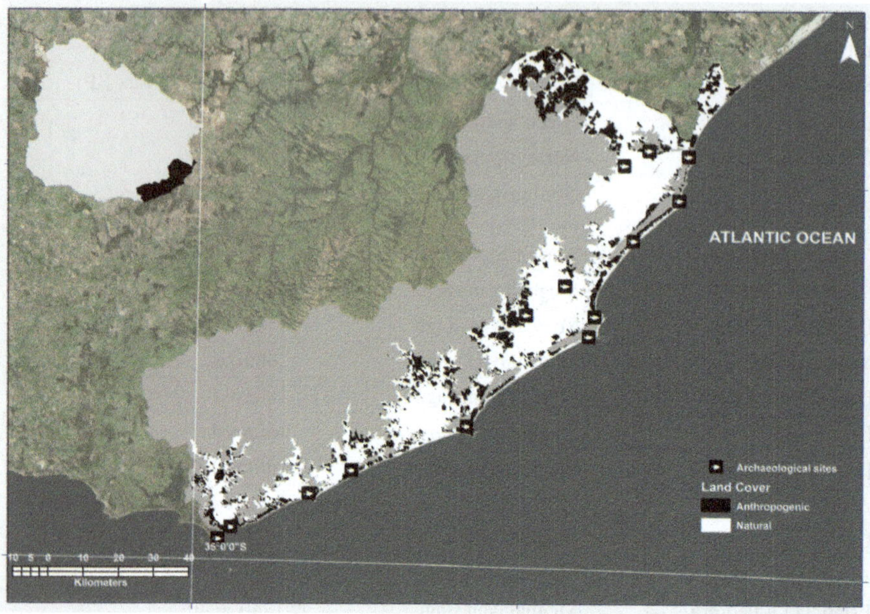

Fig. 6 Archaeological sites location in relationship with anthropogenic and natural land covers

Sites located in the western half of the coastline, mostly identified during the first half of the twentieth century by collectors, are currently located in anthropogenic covers corresponding to urban and forestry plantation areas. The same situation was identified for some of the capes located further to the east. The eastward direction gradient from anthropogenic to natural covers also implied that most Meghalayan coastal sites under natural covers were located in the eastern half of the study area. Sites located in the inland end of the study area were located under a patchy matrix that combined a growing cropland cover with natural grasslands.

Visibility and Preservation Potential of the Archaeological Record

The preceding information allowed us to evaluate the visibility and preservation potential of the archaeological record (Fig. 6). Areas where anthropogenic land cover was intersected by archaeological bearing areas were considered as low visibility and high risk for preservation. On the other hand, areas where natural covers intersected archaeological bearing layers were considered as areas of high potential visibility and preservation.

In summary, excluding coastal lagoons (assuming that human occupations did not involve territories below current lagoons coastlines), 62% of the study area were identified as low risk and high preservation potential for the existing archaeological record. As it was already mentioned in the preceding section, the remaining 38% is threatened by urban expansion, forestry plantations, and croplands. Most urban areas are located in the western portion of the study area; meanwhile, the eastern coast biggest threat is attributable to forest plantations.

Discussion and Conclusions

When available archaeological information was arranged using Holocene stages, it became apparent that coastal human occupation evidences are biased toward the Meghalayan stage. This fact is tempting enough to make the direct assumption that the Atlantic coast has only been inhabited recently in Holocene terms if coastal evolution is disregarded. Nevertheless, the strategy of this project succeeded to identify most probably coastal configurations for the different Holocene stages covered in this contribution and those scenarios depicted an entirely different panorama. The lack of archaeological evidences for the Greenlandian stage needs to be relatively considering the inferred coastline for this stage, which implied that earlier occupations should be currently underwater. Moreover, human presence was identified in the Atlantic basin at Los Indios site, an inland plain developed around a granitic outcrop that promoted the resilience of this place (and the preservation of the archaeological record) over geomorphologic processes developed through the Holocene. However, it must be noted that this site, with sea level 50 m below current coastline, was far from being a coastal occupation. These findings are in agreement with the chronologies reported for the Atlantic coast at a continental scale, where eustasy is the prevailing modeling agent across the Holocene, as Greenlandian chronologies for coastal occupations are more conspicuous in isostasy dominated shorelines (Bailey and Flemming 2008; Erlandson et al. 2008; Faught and Gusick 2011).

Archaeological evidences for the Northgrippian stage were also scarce. In this case, the explanation could be found in the combination of several factors. First, during this stage, the sea level continued the rising trend at a quick pace reaching 5 m above current sea level around 5.5 ka BP. This fact could have implied that earlier occupations were drowned and thus became obliterated by the inland displacement of the coastline. Second, identified coastal sites bearing chronologies grouped in the last millennia of this stage were located at stable landforms, such as oceanic capes and lagoon coasts. Third, the former sentence could be also shedding light over subsistence-related decisions regarding site allocation, with capes functioning as strategic base camps for opportunistic fishing or mollusk gathering on a seasonal basis. Besides, the irruption of the "Moundbuilder" groups in nearby areas during the late Northgrippian stage, in a freshwater wetland territory, could be suggesting that the coast was part of a broader territory in subsistence strategy terms (Bracco et al. 2005; Iriarte et al. 2017). These inferences are in agreement with current knowledge about subsistence strategies developed by aboriginal groups around the globe, where coastal environments are occupied and exploited as part of a broader territory, generally involving inland expansion and or interactions (Barceló et al. 2015; Staller 2001).

The climatic deterioration that defined the onset of the Meghalayan stage and triggered the collapse of several civilizations around the globe (Walker et al. 2018) has also been identified in the Atlantic coast of Uruguay and nearby areas (del Puerto et al. 2013, Inda et al. 2016). The local expression of this event implied the establishment of a colder subhumid climate with a strong seasonality in rainfall. This scenario, coupled with a decreasing sea, promoted the formation of extensive dune fields and their aeolian displacement to the northeast, invading *terra firme* territories (Panario and Piñeiro 1997). During this episode, human evidence disappeared from the coast, reflecting the abandonment of a harsh environment though the incidence of aeolian erosion should not be discarded.

As soon as the climate shifted from temperate to warm and humid and the coastal environments acquired their current configuration since 3.0 ka BP, the Atlantic coast began to be inhabited by human groups until the arrival of European conquerors. These coastal settlements comprised of not only the reoccupation of capes and coastal lagoons but also the establishment in beach arcs, interdune ephemeral ponds and after berm wetland areas (Inda et al. 2017).

When contrasted with soil information for the area, it became apparent that the archaeological record is not exclusively a consequence of the human presence during the Holocene. Some stages and episodes where cultural evidence was scarce, could be better understood by considering coastal resources availability and accessibility and it should be stressed the role of post-deposition processes (notably aeolian erosion) that could have led to the disappearance of soil units that contained cultural remains (Favier Dubois 2003).

Current land cover in the basin showed a dominance of natural grasslands, followed by riparian forest and exotic tree plantations. Nevertheless, when land covers were intersected with potential archaeological areas, anthropogenic covers, particularly urban areas and forestry plantations, exhibited higher percentages. In terms of visibility and preservation potential, the relationship between the impact of anthropogenic and natural covers is by no means as direct as it could appear. While urban areas can be easily considered as the cause behind the disappearance of several sites identified in the first half of the twentieth century, negatively affecting visibility and preservation, forestry plantations could play a different role as this cover impacted surface sites, but in some cases, stratified sites could be well preserved. The impact of cropland areas is another example of this complexity, because the impact of modern sowing techniques (particularly in the case of soybean crops) is restricted to the surface horizon.

Natural covers, on the other hand, are also a mixed bag, and they should not be directly related to positive effects on visibility and preservation of the archaeological record. Dune field cover, for example, a unit that contained most coastal sites, has a positive effect on preservation of buried sites (and a negative effect on visibility). When aeolian transport took place during storms, buried sites became exposed (positive effect on visibility) and negatively impacted by weathering and sand erosion over cultural remains.

Finally, as a basin-wide approach, this contribution is far from exhaustive as some operational decisions had to be faced where the whole picture jeopardized a higher level of detail. Nevertheless, some relevant conclusions were driven.

The Atlantic basin is experiencing widespread anthropogenic impacts that were reflected by land covers. Such an impact has an eastward gradient, so archaeological survey efforts should be allocated in the eastern portion of the basin.

Land covers presented here must be seeing as a snapshot of dynamic entities, in a recent study developed in a nearby basin, some trends in land cover changes for coastal scenarios were identified (Inda et al. 2016). These trends showed that market-driven covers, particularly real estate development and croplands, will further extend into our study area in the short term, so the archaeological record will urgently need the development of protection strategies and actions to face this imminent threat.

Acknowledgements ANII-SNI and the PDT from the Universidad de la República del Uruguay provided financial support. Authors wish to thanks GEGAL members for the prolific exchange during the Workshop held at La Paloma beach, Uruguay.

References

Angulo RJ, Lessa GC, de Souza MC (2006) A critical review of mid-to late Holocene sea-level fluctuations on the eastern Brazilian coastline. Quat Sci Rev 25:486–506

Arredondo H (1958) Viajeros visitantes del Uruguay. Revista Sociedad de Amigos de la Arqueología 15:7–313, Montevideo

Bailey GN, Flemming NC (2008) Archaeology of the continental shelf: Marine resources, submerged landscapes and underwater archaeology. Quat Sci Rev 27(23–24):2153–2165

Barceló JA, Del Castillo F, Del Olmo R, Mameli L, Miguel F, Poza D, Vilá X (2015) Simulating Patagonian Territoriality in prehistory: space, frontiers and networks among hunter-gatherers. In: Wurzer G, Kowarik K, Reschreiter H (eds) Agent-based modeling and simulation in archaeology. Advances in geographic information science. Springer

Bracco R, Inda H, del Puerto L, Capdepont I, Panario D, Castiñeira C, García Rodríguez F (2014) A reply to "relative sea level during the Holocene in Uruguay". Palaepgeography, Palaeoclimatology, Palaeoecology 401:166–170

Bracco R, García-Rodríguez F, Inda H, del Puerto L, Castiñeira C, Panario D (2011) Niveles Relativos del Mar Durante el Pleistoceno final—Holoceno en la Costa del Uruguay. In: García-Rodríguez F (ed), El Holoceno en la Zona Costera del Uruguay, 65–94. Montevideo, UCUR-Universidad de la República

Bracco R, del Puerto L, Inda H, Castiñeira C (2005) Middle-late Holocene cultural and environmental dynamics in the east of Uruguay. Quat Int 132:37–45

Bradje T, Erlandson J (2013) Looking forward, looking back: humans, anthropogenic change, and the Anthropocene. Anthropocene 4:116–121

Cavallotto J, Violante R, Parker G (2004) Sea-level fluctuations during the last 8600 years in the de la Plata river (Argentina). Quat Int 114:155–165

del Puerto L, Bracco R, Inda H, Gutiérrez O, Panario D, García-Rodríguez F (2013) Assessing links between late Holocene climate change and paleolimnological development of Peña Lagoon using opal phytoliths, physical and geochemical proxies. Quat Int 287:89–100

del Puerto L, Inda H, García-Rodriguez F, Castiñeira C, Bracco R (2006) Paleolimnological evidence of Holocene paleoclimatic changes in Lake Blanca, Southern Uruguay. J Paleolimnol 36:151–163
Durán A (1985) Los Suelos del Uruguay. Editorial Hemisferio Sur, Montevideo
Erlandson JM, Moss ML, Des Lauriers M (2008) Life on the edge: early maritime cultures of the Pacific coast of North America. Quat Sci Rev 27:2232–2245
Faught MK, Gusick A (2011) Submerged prehistory in the Americas. In: Benjamin J, Bonsall C, Pickard C, Fisher A (eds) Submerged prehistory. Ox-bow Books, Oxford, pp 145–157
Favier Dubois C (2003) Late Holocene climatic fluctuations and soil genesis in southern Patagonia: effects on the archaeological record. J Archaeol Sci 30(2003):1657–1664
Garreaud RD, Vuille M, Compagnucci R, Marengo J (2009) Present-day South American climate. Palaeogeogr Palaeoclimatol Palaeoecol 281:180–195
Gianotti C, Gascue A, del Puerto L, Inda H, Villarmarzo E (2017) Archaeological heritage on the Atlantic coast of Uruguay: heritage policies and challenges for its management in coastal protected areas. In: Dawson T, Nimura C, López-Romero E, Daire M (eds) Public archaeology and climate change (Chap 16: 149–161). Oxbow Books
Inda H, del Puerto L, Castiñeira C, Capdepont I, García-Rodríguez F (2006a) Manejo prehistórico de recursos costeros en el litoral atlántico uruguayo. In: Menafra R, Rodríguez L, Scarabino F, Conde D (eds) Bases para el Manejo y Conservación de la Costa Uruguaya. Vida Silvestre, Montevideo, pp 661–667
Inda H, García-Rodríguez F, del Puerto L et al (2006b) Relationships between Trophic State, paleosalinity and climatic changes during the first Holocene marine transgression in Rocha Lagoon, southern Uruguay. J Paleolimnol 35:699–713
Inda H, del Puerto L, Bracco R, Castiñeira C, Capdepont I, Gascue A, Baeza J (2011) Relación hombre-ambiente para la costa estuarina y oceánica de Uruguay durante el Holoceno. Reflexiones y perspectivas. In: García-Rodríguez F (ed) El Holoceno en la zona costera de Uruguay. CSIC, Universidad de la República, Montevideo, pp 229–260
Inda H, García Rodríguez F, del Puerto L, Stutz S, Lopes Figueira R, de LIma Ferreira P, Mazzeo N (2016) Discriminating between natural and human-induced shifts in a shallow coastal lagoon: a multidisciplinary approach. Anthropocene 16:1–15
Inda H, del Puerto L, Capdepont I, Bracco R (2017) Formation processes of coastal archaeological sites: a changing prehistoric scenario on the Atlantic shore of Uruguay. Geoarchaeology. 32:633–645
Iriarte J (2006) Vegetation and climate change since 14,810 ^{14}C yr B.P. in southeastern Uruguay and implications for the rise of early Formative societies. Quat Res 65:20–32
Iriarte J, De Blasis P, De Souza J, Corteletti R (2017) Emergent complexity, changing landscapes, and spheres of interaction in southeastern south America during the middle and late Holocene. J Archaeol Res 25:251–313
Isla FH (1989) Holocene sea-level fluctuations in the southern hemisphere. Quat Sci Rev 8:359–368
Isla FI, Angulo R (2016) Tectonic processes along the South America coastline derived from Quaternary marine terraces. J Coast Res 32(4):840–852
Kellog JN, Mohriak WU (2001) The tectonic and geological environment of coastal South America. In: Seeliger U, Kjerfve B (eds) Coastal marine ecosystems of Latin America. Ecological studies, vol 144, pp 1–16. Springer
López Mazz J (2013) Early human occupation of Uruguay: radiocarbon database and archaeological implications. Quat Int 301:94–103
Maeso C (1977) Investigaciones arqueológicas. In: Tuya L (compilator), Imprenta Don Bosco, Montevideo
Martin L, Suguio K (1992) Variation of coastal dynamics during the last 7000 years recorded in beach ridge plains associated with river mouths: example from the Central Brazilian Coast. Palaeogeogr Palaeoclimatol Palaeoecol 99:119–140
Martínez S, Rojas A (2013) Relative sea level during the Holocene in Uruguay. Palaeogeogr Palaeoclimatol Palaeoecol 374:123–131

Milheira RG, Gianotti García C (2018) The Earthen Mounds (Cerritos) of Southern Brazil and Uruguay. In: Smith C (ed) Encyclopedia of global archaeology. Springer

Panario D, Gutiérrez O (1999) The continental Uruguayan Cenozoic: an overview. Quat Int 62(1):75–84

Panario D, Piñeiro G (1997) Vulnerability of oceanic dune systems under wind pattern change scenarios in Uruguay. Climate Res 9:67–72

Panario D, Gutiérrez O, Sánchez Bettucci L, Peel E, Oyhantçabal P, Rabassa J (2014) Ancient landscapes of Uruguay. In: Rabassa J, Ollier C (eds) Gondwana landscapes in southern South America. Springer earth system sciences. Springer, Dordrecht

Rosa M, Barboza E, Dillenburg S, Tomazelli L, Ayup-Zouain R (2011) The Rio Grande do Sul (southern Brazil) shoreline behavior during the Quaternary: a cyclostratigraphic analysis. J Coast Res, SI 64 (Proceedings of the 11th international coastal symposium), 686–690. Szczecin, Poland

Staller JE (2001) Reassessing the developmental and chronological relationships of the Formative of coastal Ecuador. J World Prehist 15(2):193–256

Walker M, Head M, Berkelhammer M, Björck S, Cheng H, Cwynar L, Fisher D, Gkinis V, Long A, Lowe J, Newnham R, Olander S, Weiss H (2018) Formal ratification of the subdivision of the Holocene Series/Epoch (Quaternary System/Period): two new Global Boundary Stratotype Sections and Points (GSSPs) and three new stages/ subseries. Episodes, IUGS. https://doi.org/10.18814/epiiugs/2018/018016

Watanabe Y, Kawamura T, Yamashita Y (2018) Introduction: the coastal ecosystem complex as a unit of structure and function of biological productivity in coastal areas. Fish Sci 84:149

Werz B, Cawthra H, Compton J (2014) Recent developments in african offshore prehistoric archaeological research, with an emphasis on South Africa. In: Evans A, Flatman J, Flemming N (eds) Prehistoric archaeology on the continental shelf. Springer, New York

OSL Dating of Lagoon Geoforms as Proxies of Marine Levels for the Late Holocene

Daniel Panario, Roberto Bracco Boksar, Ofelia Gutiérrez and Marcos Tassano

Abstract The Castillos Lagoon has been a research focus regarding sea level fluctuations and associated climatic changes during the Holocene due to its well-preserved record of geomorphological, stratigraphic, and geoarchaeological structures. In this paper, two series of OSL (Optically Stimulated Luminescence) datings will be presented. The first one comes from a system of cheniers situated north of the Castillos Lagoon. The data obtained show that at least during a 1,200 year period, counting from a few years before 2460 yr BP, there was a process of progradation of the lagoon coast and the development of a chenier plain, consistent with a gradual drop in sea level, within a scenario of relatively stable conditions. The second series, proceeding from the first beach ridge adjacent to the lagoon coast, has allowed to determine that this stripe is current and therefore still under formation, or recently formed, permitting to obtain a direct measurement of the difference between its height and the lagoon level related to the moment of its formation that may be correlated to analogous historical records.

Keywords Chenier · Sea levels · Coastal lagoons · OSL · Holocene · Uruguay

---

D. Panario · O. Gutiérrez (✉)
UNCIEP, Instituto de Ecología y Ciencias Ambientales (IECA), Facultad de Ciencias, Universidad de la República, Iguá 4225, 11400 Montevideo, Uruguay
e-mail: oguti@fcien.edu.uy

R. Bracco Boksar
Facultad de Humanidades y Ciencias de la Educación, Universidad de la República, Av. Uruguay 1695, 11200 Montevideo, Uruguay

Laboratorio 14C, Facultad de Química, Universidad de la República - Ministerio de Educación y Cultura, General Flores 2124, 11200 Montevideo, Uruguay

CURE, Sede Rocha, Universidad de la República, Route No. 9 Km 207, 27000 Rocha, Uruguay

M. Tassano
Laboratorio de Radioquímica, Centro de Investigaciones Nucleares, Facultad de Ciencias, Universidad de la República, Iguá 4225, 11400 Montevideo, Uruguay

H. Inda Ferrero and F. García Rodríguez (eds.), *Advances in Coastal Geoarchaeology in Latin America*, The Latin American Studies Book Series,
https://doi.org/10.1007/978-3-030-17828-4_3

Introduction

Coastal lagoons are areas of high environmental memory since their genesis and geomorphological history are determined by geological, climatic, hydrological and ecological changes, as well as changes in sea level (Barnes 1980; Bird 1994; Martin and Landim Dominguez 1994).

The Castillos Lagoon (34°19′17.49″S–53°55′41.97″W) is a brackish lagoon with a basin area of 80 km^2 (Panario and Gutiérrez 2011). It has maintained an active connection with the ocean throughout the Holocene (Bracco et al. 2011a; Inda et al. 2017). The Valizas stream is the link through which, depending on rain, tides, and the state of its bar, it drains or receives seawater, forming an estuarine system (Conde et al. 2003). The basin has an area of 1,453 km^2 (García-Rodríguez et al. 2011). Different environments showing an estuarine regime are found in this basin: mountain ranges, hills, high, middle and low plains, permanent lagoons, seasonal wetlands, coastal plains, dune fields, and oceanic beaches. Its principal tributaries are the Don Carlos, Chafalote, del Consejo, Sarandí, and Castillos streams.

The Castillos Lagoon forms part of a system of coastal lagoons that lies in eastern Uruguay and continues through the southern coast of Brazil (Panario and Gutiérrez 2011). It lies south of the Merín Lagoon Basin where the climate is of subtropical temperate nature with a mean rainfall of 1326 mm (1980/2009 period, La Paloma weather station) and an annual mean temperature of 16.8 °C (Rocha weather station) (Castaño et al. 2011).

The Castillos Lagoon was formed in a small subbasin of the Pelotas Sedimentary Basin (Sprechmann 1980), within the Crystalline Basement (540–520 Ma). In its margins, it is possible to observe Quaternary sedimentary deposits, located in elevation levels placed below those where the pre-Pleistocene rocky formations emerge. The oldest ones correspond to the Libertad and Lomas de Narváez formations, both of Pleistocene age (Preciozzi et al. 1985), being the latter probably originated as a coastal barrier during Pleistocene transgressions (Villwock et al. 1986; Bracco et al. 2011b), which gives evidence of a cyclical process. Below the Quaternary deposits between the level of 10 and 5 m above sea level (m a.s.l. hereafter) lies the Dolores Formation, which appears as a flat surface composed of friable silt deposits. It is assigned to the final Pleistocene (Panario and Gutiérrez 2011). Below 5 m a.s.l., there are Holocene deposits of the Villa Soriano Formation (Preciozzi et al. 1985). In the inner lagoon littoral, there appear various landforms associated to the Holocene fluctuations of the lagoon level (Fig. 1), which—observing the connection to the ocean—are correlated with the regional fluctuations in sea level. The most conspicuous ones are erosion terraces, tidal tide plains, barrier systems, beach ridges, cheniers, and coastal deposits (Bracco et al. 2011b; Panario and Gutiérrez 2011).

The Castillos Lagoon became the focus of research related to Holocene sea level changes within the system of coastal lagoons in Uruguay due to its well-preserved record, including a high number of geomorphological, stratigraphic, and geoarchaeological structures. From it, focusing on the altimetry and ^{14}C chronologies of the

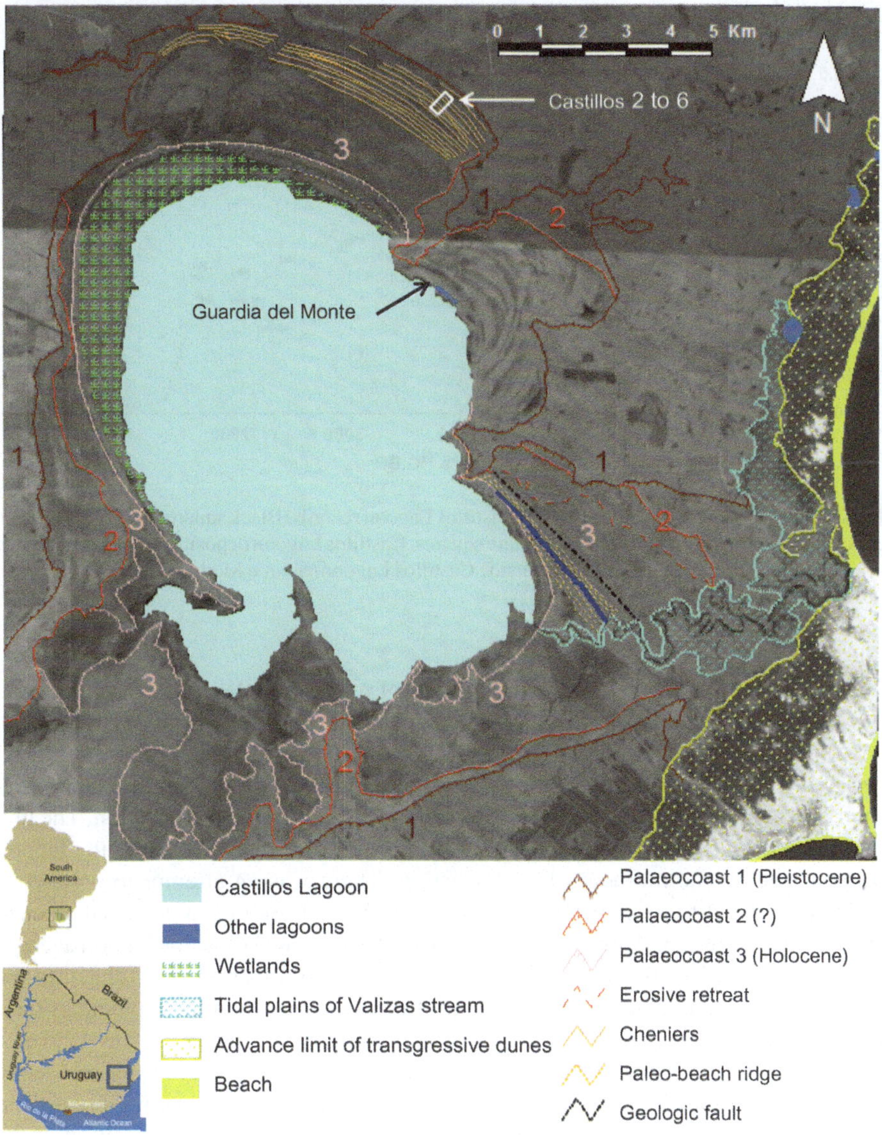

Fig. 1 Geomorphological map of the Castillos Lagoon. Arrows indicate sampled localities for OSL dates: Castillos Lagoon and Guardia del Monte

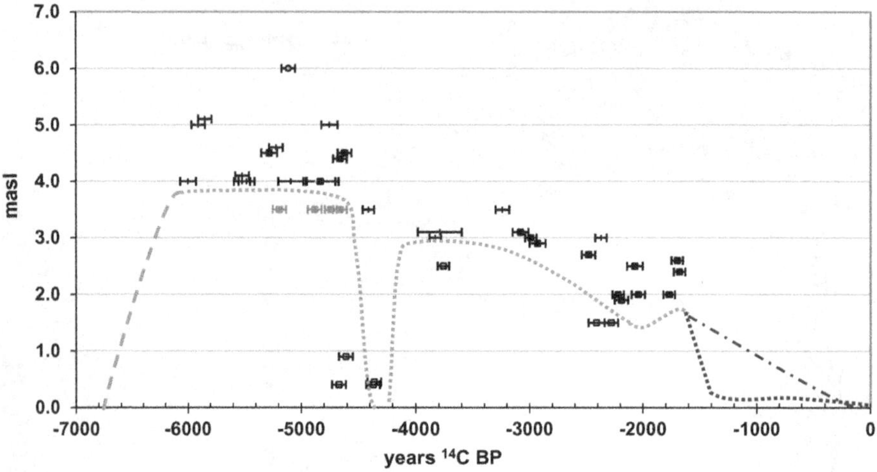

Fig. 2 The sea level curve inferred from Castillos Lagoon records. Black squares: Castillos Lagoon deposits formed above mean sea level. White squares, Castillos Lagoon deposits formed below mean sea level. Gray circles: samples taken outside Castillos Lagoon. *Source* Modified from Bracco et al. (2014)

geoforms found in its coasts, the first curves of relative sea level for Uruguay were developed (Bracco and Ures 1998; Bracco et al. 2011b; Bracco et al. 2014) (Fig. 2).

Although this meant significant progress in determining the trend of sea level behavior in this area, these models are based mainly on two assumptions that limit their validity in order to achieve a greater altimetric and temporal resolution. The first one refers to the ratio between the elevation of the dated geoforms and the lagoon normal level at the moment of its formation. There was an attempt to precise an estimate of this difference through the articulation of chronological data that proceed from geoforms and/or deposits that were formed above and below the lagoon average level (see Fig. 13 in Bracco et al. 2011b). The second assumption is related to the time between the age of the dated samples (marine valves) and the moment of their deposition. In order to specify it, particular attention was devoted to any evidence of transportation in the dated valves, but it should not be ignored that moderate presence or absence of evidence of transport does not allow to conclusively uphold the contemporaneity between the age of the dated material and its deposition. If the character of parautochthonous (sensu Kidwell et al. 1986, p. 229) is assigned to the context of provenance of the valves, the opposite is acknowledged (Bracco et al. 2011b, p. 82). Furthermore, the possible presence of reservoir effect should be added (Stuiver and Braziunas 1993).

In this paper, two series of Optically Stimulated Luminescence (hereafter OSL) datings are presented. The first one comes from a system of cheniers located in the northern shore of the Lagoon. The second comes from the first beach ridge adjacent to the current lagoon coast. As the OSL record is initiated at once when the sediments are buried, OSL dating defines a direct age of deposit formation. By determining that the first ridge is modern, and therefore, it is either under formation or has been recently formed, the difference between its height—at the time of its formation—and the lagoon level may be measured. This is a direct measurement which can be extrapolated to analogous historical records.

Archaeological Background

In the surroundings of the Castillos Lagoon, 23 archaeological sites have been identified, containing one, two, or three mounds (Pintos Blanco 1999; Pintos Blanco 2001; Pintos Blanco and Capdepont 2001; López and Gascue 2007; Bracco et al. 2015). The height of these mounds does not exceed 1.2 meters. These sites are located predominantly between elevations of 5 and 20 m a.s.l., on gentle slopes modeled on the Libertad and Dolores Formations.

Cráneo Marcado and Guardia del Monte have been the only two sites systematically excavated (Castiñeira et al. 1999; Capdepont 2001; Capdepont and Pintos Blanco 2002).

For the first site, there are two ^{14}C dates made on charcoal. One of them comes from a stratigraphic unit with cultural evidence (UE II) that lies at a coastline (beach ridge): 3050 $\pm$ 150 ^{14}C yr BP (Pintos Blanco and Bracco 1999, p. 89). The second dating comes from EU II and III of Monticular Structure B: 2760 $\pm$ 60 ^{14}C yr BP (Capdepont and Pintos Blanco 2006, Table 1). For the Guardia del Monte site there are also two radiocarbon dates: 1260 $\pm$ 60 ^{14}C yr BP and 4600 $\pm$ 60 ^{14}C yr BP. The first dating was made on charcoal from EU II of mound A. The second was made on shells, and it is not directly associated with cultural materials, thus it only allows to indicate a maximum age of occupation of the site (see Inda et al. 2017, Fig. 5). Human burials were found on mound B of Cráneo Marcado and mound A of Guardia del Monte. Capdepont and Pintos Blanco (2002) reported the presence of at least 10 individuals, and they recorded primary, secondary, individual, and collective burials.

Culverin bullets (ancestor of the musket used in the sixteenth and seventeenth centuries) and fragments of a porcelain mold pipe were identified as the grave goods of one of the burials, which indicated that their chronology of the burials reaches the period of European contact (Pintos Blanco and Capdepont 2001). The recovered lithic artifacts are composed of "boleadoras" (balls), scrappers, nut crackers, lithic cores, pestles, and mortars, made mostly from local raw materials (Pintos Blanco and Capdepont 2001). The recovered ceramic corresponds to utilitarian type vessels. For both sites, Capdepont and Pintos Blanco (2006) indicated that the hunting activities focused on continental resources, mainly cervids, although aquatic species on the lagoon and Atlantic littoral are also represented. Analyses of biosiliceous particles and starch granules made from sediments, sherds, and grinding artifacts indicate the use, processing, and/or storage of wild and domesticated plants. The record includes palm trees, Cannaceae, Cyperaceae, Bromeliaceae, grasses of the Oryzeae and Bambuseae tribes, Equisetaceae, Cannabaceae, and Orieceae grasses, along with maize (*Zea mays*), beans (*Phaseolus*), and pumpkins (Cucurbitaceae) (del Puerto et al. 2016, Fig. 14). Del Puerto and collaborators pointed to an intensification in the exploitation of resources, from ~2000 to 1500 cal. years BP, being corn and beans—domesticated resources—the last to join the record.

Materials and Methods

OSL Dating

Minerals like quartz are very suitable for Optically Stimulated Luminescence dating since they have the property of accumulating background radiation (ionizing energy). The amount of energy accumulated (paleodose) is proportional to the intensity (radiation unit/time) and the exposure time. If these minerals are exposed to light, the accumulated energy is liberated in the form of light, resetting the luminescence "clock" to zero (bleached). If these minerals are later protected from light (buried) the cycle restarts. By measuring the paleodose and the annual dose (amount of background radiation/year), the time elapsed since the mineral last exposure to light or, equally, since it was buried, may be measured. Age: paleodose/annual dose (Aitken 1974, 1985, 1997, 1998; Murray and Olley 2002; Shrestha 2013).

Sampling

Sampling involved two areas of the north-northeastern sector of the Lagoon. In both cases, test pits were excavated and profile interpretation was performed, as well as sample gathering for OSL dating. In order to avoid samples to be directly exposed to sunlight, opaque polyethylene tubes were horizontally inserted at least 30 cm from the profile exposed surface, and then extracted samples were immediately wrapped with aluminum foil. The first sampled area was the beach ridge next to the current coastline. This beach ridge has a maximum height of 60 cm over the lagoon level, and its width is approximately 40 m. The three recognized stratigraphic units were sampled for dating (Fig. 3). The second area is located in an old inlet lagoon present between 2.0 and 1.5 m a.s.l., and its northern boundary (outer) is the erosion terrace attributed to the Holocene Maximum Transgression, carved on the Dolores Formation. Attached to this formation, a sandy bar extends to the west where three calibrated ^{14}C datings were obtained: 5190 ± 60 ^{14}C yr BP (Cal. BP (2σ): 5724–6018 ($p = 0.974$), 6080–6108 ($p = 0.017$), and 6157–6172 ($p = 0.008$), 4880 ± 60 ^{14}C yr BP (Cal. BP (2σ): 5331–5374 ($p = 0.045$) and 5459–5715 ($p = 0.995$), and 4740 ± 50 ^{14}C yr BP (Cal. BP (2σ): 5316–5487 ($p = 0.754$) and 5506–5582 ($p = 0.246$) (Bracco et al. 2014). The calibration of the conventional radiocarbon ages was obtained using ShCal13 curve for the Southern Hemisphere (Hogg et al. 2013). In the inlet, there is a group of parallel geoforms, in the shape of a wide open arch, approximately 5000

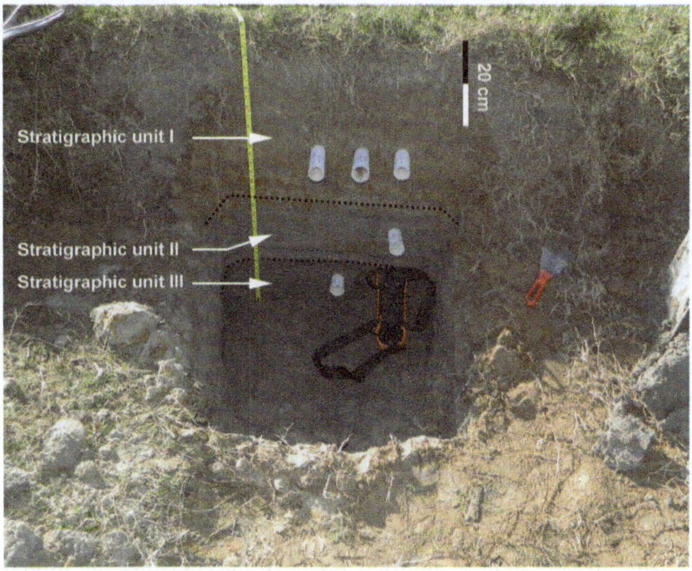

Fig. 3 Stratigraphic units of the beach ridge near the current Castillos Lagoon coastline and sampling tubes for OSL

m long and 70 m wide each (see Fig. 1). Four adjacent geoforms were sampled in this area with 60 cm deep sampling holes. Samples for dating were extracted at a depth of 40 cm. In the exposed profiles only two stratigraphic units can be observed. In the upper unit, fine to very fine sand dominates, and only the horizons of current soil were identified. The lack of other soil structures in the sampled profiles is attributed to edaphic processes that erased formed soil developments. The lower one, below 50 cm deep, with finer grain size (fine sand and silt), showed clear evidence of hydromorphism (grey colors).

Age Estimation

The OSL datings were carried out in the Laboratory of Luminescence at the Facultad de Ciencias de la Universidad de la República (Montevideo, Uruguay). First, under suitable artificial light, one centimeter of sediment from each face of the sample tube was eliminated. The remaining sediment was deflocculated using sodium hexametaphosphate (5%), organic matter was eliminated with Hydrogen Peroxide and the carbonates with 8% of HCl. Sediment fractions were separated by sieving with 63–90 μm and 90–150 meshes.

Both fractions were attacked with Fluoride acid 30% during one hour in order to eliminate feldspar and the outer core of the quartz grains. Through successive rinsing in distilled water, it was brought to neutral pH. After that, HCI at 38% was added during 10 min to remove the fluoride produced in the previous stage. The residue was oven dried at 40 °C for 48 h. Finally, a double separation using LST (™) with densities of 2.75 and 2.63 g/ml was carried out. After the drying process, the measurement aliquots were mounted on 1 cm diameter stainless steel disks using a silicon-based synthetic adhesive. The OSL readings were carried out with a Daybreak™ 1100 device. Daybreak™ model 801 E was used for the irradiation, equipped with a beta ^{90}Sr source. The initial preheating temperature was determined from the intersection point of the TL irradiated and natural (additive) spectra. 10 °C increases compared TL and OSL spectra of natural and irradiated aliquots at successive intervals. The temperature at which the TL and OSL spectra normalized from the natural and irradiated aliquots appeared as equal was selected. Additive and regenerative methods of multiple aliquots (MARD and MAAD) were followed for the measurement (Vandenberghe 2004). A previous measurement of the luminescent signal of each aliquot was carried out in order to compensate differences in load and sensitivity. This was achieved by stimulating samples for 0.15 s. These measurements allowed to calculate a correction factor. When following the MARD method, bleaching was carried out by exposing the aliquots to sunlight for 60 min, checking the effectiveness of the bleaching by measuring three aliquots. A minimum of 10 natural aliquots was measured, and 6 for every one of the 4 or 5 steps of irradiation. It was only accepted as valid the series of natural measurements or of each irradiation step that presented CV $\leq$5%, discarding only a maximum of 2 aliquots per series. Otherwise, the series was repeated. The concentrations of ^{238}U, ^{232}Th, ^{40}K, and ^{137}Cs were determined with a gamma spectrometer (Canberra of hyperpure germanium type-P, with relative efficiency: >20% to 1.33 MeV) in the Laboratory of Radiochemistry at the Center of Nuclear Research of the Universidad de la República (Montevideo, Uruguay). In order to calibrate the equipment, a soil standard was used, certified by the International Atomic Energy Agency model IAEA-447 which has multiple natural and artificial radionuclides (Table 1). The age calculations were made following Adamiec and Aitken (1998).

Results and Discussion

Table 2 shows the 7 OSL ages obtained from the lagoon current ridge, and Table 3 shows the 13 OSL ages obtained from the four dated cheniers. In the case of the ridge, the ages corresponding to the second stratigraphic unit spanned over the 1922–1945 period. They are statistically non-distinguishable, and since they have the same origin, it can be considered their weighted mean and weighted error as the best estimation of the age of this stratigraphic unit: 81 $\pm$ 17 years (1935 $\pm$ 17 AD). If two sigmas are considered, there is a 97% probability that the real age is between the years 1969 and 1901 of this era. The ages which correspond to the third stratigraphic unit cover the 1920–1957 period. Its weighted mean is 76 $\pm$ 21 years. If two sigmas are considered, there is a 97% probability that the real age is in the period between the years 1982

Table 1 Concentrations of ^{238}U, ^{232}Th, ^{40}K, and ^{137}Cs from the sediments sampled

Sample	^{238}U (ppm)	Error (%)	^{232}Th (ppm)	Error (%)	K (%)	Error (%)	^{137}Cs (Bq/Kg) Activity as of 01/01/2014	Error (%)
Castillos	1.95	8.26	9.60	8.92	4.33	5.85		
Guardia del Monte 2 (II)	0.96	8.89	3.79	9.70	2.95	5.97		
Guardia del Monte 2 (III)	1.26	8.60	4.85	9.45	3.88	5.91	2.45	7.9
Guardia del Monte 2 (IV)	2.32	8.14	9.51	8.95	4.75	5.83		

and 1898 AD. At this level, the gamma spectrometry detected ^{137}Cs (Table 1). This artificial isotope was introduced into the atmosphere as a consequence of nuclear tests. Its integration to soils starts shortly after nuclear test derivatives fallout in 1954 and ends in the 1980 decade, reaching its peak in 1963 (Ritchie and Ritchie 1995). The fourth stratigraphic unit that is located below the ridge has two dates. Their weighted mean is 132 ± 28 years. If two sigmas are considered, it may be sustained with a 97% probability that the real age is between the years 1940 and 1828 of this era. The set of dates narrow down the formation process of this geoform to a very recent period, and it cannot be discarded discard that it still continues to develop.

Thirteen OSL ages are obtained from the plain that lies in the northern part of the inlet lagoon. These geoforms were acknowledged by Panario and Gutiérrez (2011) as a chenier plain, following Otvos and Price (1979) and Augustinus (1989). Cheniers are ridges formed by siliciclastic and bioclastic materials that are generally situated in the highest part of the tidal plains or supratidal plains. They are produced by the reworking of sediments (sand and shell fragment) by the waves which transported and deposited the sediment over the muddy flats along coastlines and forming small ridge. Otvos and Price (1979) proposed that a chenier plain is present when it may be identified at least two of these parallel sandy ridges separated by mudflats. Bracco and colleagues inferred such arrangement as a system of ridges (see Fig. 10 in Bracco et al. 2011b). At any rate, they are acknowledged as wave-built geoforms at the highest end of tidal plains or supratidal plains. In our case, the presence of mudflats below and between the geoforms, as well as the predominance of fine sands in their sediment, would indicate that their formation took place in low energy environments, which is the referred case for chenier formation.

Table 2 OSL ages coming from the lagoon current ridge

Origin	Methodology	Gy	σ	Age OSL years	Age OSL years AP	σ	M_p	σ_p
G. Monte 2-2	AD	0.050	0.020	70	5	30	15	17
G. Monte 2-2	AD	0.070	0.007	93	28	30		
G. Monte 2-2	REG	0.060	0.006	80	15	30		
G. Monte 2-3	REG	0.050	0.005	58	−7	30	11	21
G. Monte 2-3	AD	0.080	0.008	95	30	30		
G. Monte 2-4	REG	0.190	0.019	130	65	40	67	28
G. Monte 2-4	REG	0.190	0.019	135	70	40		

Table 3 OSL ages coming from the four dated cheniers

Origin	Methodology	Gy	σ	Age OSL years	Age OSL years AP	σ	M_p	σ_p
Castillos 2	REG	3.330	0.310	2430	2365	260	2395	150
Castillos 2	REG	3.590	0.3340	2615	2550	270		
Castillos 2	REG	3.240	0.303	2360	2295	250		
Castillos 3	REG	2.885	0.276	2095	2030	230	2106	130
Castillos 3	ADIT	2.987	0.286	2170	2105	230		
Castillos 3	ADIT	3.093	0.296	2250	2185	230		
Castillos 4	REG	2.012	0.193	1440	1375	168	1325	95
Castillos 4	AD	1.839	0.176	1311	1246	164		
Castillos 4	AD	2.001	0.192	1424	1359	165		
Castillos 6	AD	1.824	0.175	1300	1235	140	1150	70
Castillos 6	REG	1.662	0.159	1180	1115	135		
Castillos 6	AD	1.599	0.153	1131	1066	125		
Castillos 6	REG	1.782	0.171	1268	1203	130		

Chenier formation infers by itself a shortage of coarse- to medium-sized sediment particles, which is attributed to a wetter climate and very stable lagoon levels (open bar). For the chenier geoform located further away from the coast, four OSL ages were obtained, and three other dating are available for the following two chenier nearest to the coast. Statistically non-differentiable data compose each one of the dating series. Their weighted mean and errors are the best estimate of the age: 2395 ± 150, 2106 ± 130, 1325 ± 95, and 1150 ± 70 yr BP, respectively (Table 2). Moreover, the better climatic conditions for this period can be correlated with the greater record of biosiliceous particles generated by agricultural practices in archaeological sites.

All of these ages are earlier than the ^{14}C ages coming from the barrier north of the lagoon inlet, which due to its geomorphological position is necessarily older. They also follow the expected chronological sequence to their relative geomorphological position. Finally, it may be stated that their altimetry and chronology are consistent with the relative sea level curve proposed for the coasts of Uruguay based upon ^{14}C ages (see Fig. 2).

It is possible to hypothesize that the mouth of the lagoon must have been entirely open during the processes of chenier formation, at least almost permanently. This is indicated by the persistence and conservation of chenier structures which were not destroyed or covered by deposits generated from extraordinary lagoon flood tides.

Conclusions

The first series of OSL datings presented allows to state that the closest ridge to the lagoon—which is developing or has been recently formed—was generated just a few decimetres above the current average height level of the lagoon. This offers direct data to estimate the difference between the lagoon standard height and the altimetry of similar geoforms, the age, and altimetry of which have been used to determine sea level Holocene history for the coasts of Uruguay, therefore reducing the uncertainty of the models that represent it.

The second series of data indicate that the chenier system located in the lagoon northern inlet was developed during 2395 ± 150 and 1150 ± 70 years BP. Taking into account their altimetric position and age it may be observed a high correspondence with the lagoon (sea) relative levels proposed for that time lapse by Bracco et al. (2011). This not only broadens the empirical base of the abovementioned sea level curve but also contributes with higher resolution data sets. The cheniers are formed slightly above middle sea level environments—in micro-tidal environments and, consequently, are better indicators of sea level than other sandy geoforms that generate during storm surges.

Finally, it must be stressed that the OSL technique allows obtaining direct chronologies of the sedimentary records that have been used as indicators of the sea paleo-levels for the region, overcoming the uncertainty between dating with ^{14}C between the ages of sediment and the material of which the deposits were generated, and the ages of the deposition. In the case of the Castillos Lagoon, it has also allowed to extend to modern days the time lapse until now roughly covered by ^{14}C datings.

Additionally, from a morphogenetic point of view, the second series of data shows that during at least a 1200 year period from a few years before 2400 yr BP, a process of progradation of the coast took place without significant interruptions. This process is consistent with a gradual decrease of sea level in a scenario of relatively stable conditions of biostasy sensu Erhart (1956). It should be noted that during the period

3500–2400 yr BP the bottom deposits are almost absent from the coastal lagoons in Uruguay (see Bracco et al. 2011b; del Puerto 2015).

Finally, it must be highlighted that OSL dating is very useful to calibrate the Holocene curves of the mean sea level since it allows to obtain direct chronologies of the main records used as indicators of sea paleo-levels in the region. It also allows overcoming the uncertainty between the ages of deposits and samples, which is particularly valid for those environments with structures that, analyzed and interpreted by interdisciplinary teams, presented a set of good proxy records that may be correlated, which stimulate to continue these studies in the Atlantic Coast and the River Plate estuary.

References

Adamiec G, Aitken MJ (1998) Dose-rate conversion factors: update. Anc TL 16:37–50
Aitken MJ (1974) Physics and archaeology, 2nd edn. Clarendon Press, Oxford, England
Aitken MJ (1985) Thermoluminescence dating. Academic Press, London, England
Aitken MJ (1997) Luminescence dating. In: Taylor RE, Aitken MJ (eds) Chronometric dating in archaeology. Springer Science + Business Media, New York, USA, pp 183–216
Aitken MJ (1998) An introduction to optical dating. The dating of quaternary sediments by the use of photon-stimulated luminescence. Oxford Books, New York, USA
Augustinus PGEF (1989) Cheniers and chenier plains: a general introduction. Mar Geol 90:219–229. https://doi.org/10.1016/0025-3227(89)90126-6
Barnes RSK (1980) Coastal Lagoons. The natural history of a neglected habitat. University Press, Cambridge, United Kingdom
Bird ECF (1994) Chapter 2 Physical setting and geomorphology of coastal Lagoons. In: Kjerfve B (ed) Coastal Lagoon processes, vol 60. Elsevier Oceanography Series. Elsevier, pp 9–39
Bracco R, Ures C (1998) Las variaciones del nivel del mar y el desarrollo de las culturas prehistóricas del Uruguay. Rev do Mus Arqueol e Etnol 8:109–115
Bracco R, García-Rodríguez F, Inda H, del Puerto L, Castiñeira C, Panario D (2011a) Niveles relativos del mar durante el Pleistoceno final-Holoceno en la costa de Uruguay. In: García-Rodríguez F (ed) El Holoceno en la zona costera de Uruguay. Departamento de Publicaciones, Unidad de Comunicación de la Universidad de la República (UCUR), Montevideo, Uruguay, pp 65–92
Bracco R, del Puerto L, Inda H, Panario D, Castiñeira C, García-Rodríguez F (2011b) The relationship between emergence of mound builders in SE Uruguay and climate change inferred from opal phytolith records. Quat Int 245:62–73. https://doi.org/10.1016/j.quaint.2010.05.025
Bracco R, Inda H, del Puerto L, Capdepont I, Panario D, Castiñeira C, García-Rodríguez F (2014) A reply to 'Relative sea level during the Holocene in Uruguay'. Palaeogeogr Palaeoclimatol Palaeoecol 401:166–170. https://doi.org/10.1016/j.palaeo.2013.10.012
Bracco R, Inda H, del Puerto L (2015) Complejidad en montículos de la cuenca de la laguna Merín y análisis de redes sociales. Intersecc en Antropol 16:271–286. Olavarría, Argentina
Capdepont I (2001) Análisis cerámico en la región este del Uruguay. In: X Congreso Nacional de Arqueología Uruguaya. Arqueología uruguaya. Hacia el fin del milenio, vol 2, Colonia del Sacramento, 16–19 de Junio 1997. Gráficos del Sur, Montevideo, Uruguay, pp 377–386
Capdepont I, Pintos Blanco S (2002) Manifestaciones funerarias de los constructores de cerritos: enterramientos humanos en los túmulos de la Laguna de Castillos, Departamento de Rocha, Uruguay. In: Mazzanti D, Berón M, Oliva F (eds) Del mar a los salitrales. Diez mil años de historia pampeana en el umbral del tercer milenio. II Congreso de Arqueología de la Región Pampeana

Argentina. Universidad Nacional de Mar del Plata, Facultad de Humanidades, Laboratorio de Arqueología, Mar del Plata, Argentina, pp 107–120

Capdepont I, Pintos Blanco S (2006) Manejo y aprovechamiento del medio por parte de los grupos constructores de montículos: Cuenca de la Laguna de Castillos, Rocha–Uruguay. Relac Soc Argent Antropol 31:117–132. Buenos Aires

Castaño JP, Giménez A, Ceroni M, Furest J, Aunchayna R, Bidegain M (2011) Caracterización agroclimática del Uruguay 1980–2009 (INIA Serie Técnica, 193). Instituto de Investigaciones Agropecuarias, Montevideo, Uruguay

Castiñeira C, Fernández G, Pintos Blanco S, Piñeiro G (1999) Aplicaciones del estudio paleoambiental a las investigaciones arqueológicas. Laguna de Castillos (Rocha, Uruguay). In: Diez Marín C (ed) Actas del XII Congreso Nacional de Arqueología Argentina. Universidad Nacional de La Plata, La Plata, Argentina, pp 308–309

Conde D, Rodríguez-Gallego L, Rodríguez-Graña L (2003) Análisis conceptual de las Interacciones abióticas y biológicas entre el océano y las lagunas de la costa atlántica de Uruguay. PNUD/GEF/RLA/99/G31 (FREPLATA-Ciencias 8) Sección Limnología, Facultad de Ciencias/Universidad de la República, Montevideo, Uruguay

del Puerto L (2015) Interrelaciones humano-ambientales durante el Holoceno tardío en el este del Uruguay: Cambio Climático y Dinámica Cultural. Unpublished doctoral thesis, PEDECIBA, Universidad de la República, Montevideo, Uruguay

del Puerto L, Capdepont I, Inda H (2016) Paleoetnobotánica y subsistencia de los constructores de Cerritos del Holoceno Tardío en el Este del Uruguay: análisis fitolítico en sedimentos y artefactos arqueológicos. Tessituras: Rev Antropol e Arqueol 4:117–160. http://sci-hub.tw/10.15210/tes.v4i1.8254

Erhart H (1956) La genèse des sols en tant que phénomène géologique. Esquisse d'une théorie géologique et géochimique, biostasie et rhexistasi. Masson, Paris, France

García-Rodríguez F, del Puerto L, Bracco R, Panario D, Castiñeira C, Mourelle D, Inda H, Gordon N, Adams J (2011) Paleolimnología de las lagunas costeras del sudeste de Uruguay, y comparación paleogeográfica entre América y África del Sur. In: García-Rodríguez F (ed) El Holoceno en la zona costera de Uruguay. Departamento de Publicaciones, Unidad de Comunicación de la Universidad de la República (UCUR), Montevideo, Uruguay, pp 173–196

Hogg AG, Hua Q, Blackwell PG, Niu M, Buck CE, Guilderson TP, Heaton TJ, Palmer JG, Reimer PJ, Reimer RW, Turney CSM, Zimmerman SRH (2013) $SHCal_{13}$ Southern Hemisphere calibration, 0–50,000 cal. years BP. Radiocarb 55:1889–1903. https://doi.org/10.2458/azu_js_rc.55.16783

Inda H, del Puerto L, Capdepont I, Bracco R (2017) Formation processes of coastal archaeological sites: a changing prehistoric scenario on the Atlantic shore of Uruguay. Geoarchaeology 32:633–645. https://doi.org/10.1002/gea.21644

Kidwell SM, Fursich FT, Aigner T (1986) Conceptual framework for the analysis and classification of fossil concentrations. Palaios 1:228–238. https://doi.org/10.2307/3514687

López JM, Gascue A (2007) En el valle del Arroyo Balizas: estructuras monticulares y sitios del Litoral Atlántico Uruguayo. Cazadores-Recolectores del Cono Sur. Rev Arqueol 2:89–103. Buenos Aires

Martin L, Landim Dominguez JM (1994) Chapter 3 Geological history of coastal Lagoons. In: Kjerfve B (ed) Coastal Lagoon processes, vol 60. Elsevier Oceanography Series. Elsevier, pp 41–68

Murray AS, Olley JM (2002) Precision and accuracy in the optically stimulated luminescence dating of sedimentary quartz: a status review. Geochronometria 21:1–16

Otvos EG, Price WA (1979) Problems of chenier genesis and terminology—an overview. Mar Geol 31:251–263. https://doi.org/10.1016/0025-3227(79)90036-7

Panario D, Gutiérrez O (2011) Introducción a la geomorfología de lagunas costeras, lagos someros y charcas de Uruguay. In: García-Rodríguez F (ed) El Holoceno en la zona costera de Uruguay. Departamento de Publicaciones, Unidad de Comunicación de la Universidad de la República (UCUR), Montevideo, Uruguay, pp 49–63

Pintos Blanco S (1999) Túmulos, caciques y otras historias. Cazadores recolectores complejos en la cuenca de la laguna de Castillos, Uruguay. Complutum 10:213–226. Madrid

Pintos Blanco S (2001) Arqueología en el sitio Cráneo Marcado - Laguna de Castillos, Rocha (R.O.U.). In: X Congreso Nacional de Arqueología Uruguaya. Arqueología uruguaya. Hacia el fin del milenio, vol. I, Colonia del Sacramento, 16–19 de Junio 1997. Gráficos del Sur, Montevideo, Uruguay, pp 207–222

Pintos Blanco S, Bracco R (1999) Modalidades de enterramiento y huellas de origen antrópico en especímenes óseos humanos. Tierras Bajas del Este del Uruguay (R.O.U). In: López JM, Sans M (eds), Arqueología y bioantropología de las Tierras Bajas. Simposio 'Arqueología y Bioantropología de las Tierras Bajas,' 49° Congreso Internacional de Americanistas, Setiembre 1997, Quito, Ecuador. Universidad de la República, Facultad de Humanidades y Ciencias, Departamento de Publicaciones, Montevideo, Uruguay, pp 81–106

Pintos Blanco S, Capdepont I (2001) Arqueología en la Cuenca de la Laguna de Castillos. Apuntes sobre complejidad cultural en sociedades cazadoras-recolectoras del Este del Uruguay. ArqueoWeb—Revista sobre Arqueología en Internet 3:1–15. Madrid

Preciozzi F, Spoturno J, Heinzen W, Rossi P (1985) Memoria explicativa de la Carta Geológica del Uruguay a escala 1:500.000. Dirección Nacional de Minería y Geología, Ministerio de Industria, Energía y Minería, Montevideo, Uruguay

Ritchie JC, Ritchie CA (1995) ^{137}Cs use in erosion and sediment deposition studies: promises and problems. In: Use of nuclear techniques in studying soil erosion and siltation. Proceedings of an Advisory Group meeting held in Vienna, 26–29 April 1993. IAEA—International Atomic Energy Agency, Vienna, Austria, pp 111–124

Shrestha R (2013) Optically stimulated luminescence (OSL) dating of aeolian sediments of Skåne, southern Sweden. Unpublished MSc thesis, Lund University, Lund, Sweden

Sprechmann P (1980) Paleoecología, paleogeografía y estratigrafía de la región costera del Uruguay durante el Neógeno y Cuaternario. In: Actas II Congreso Argentino de Paleontología y Bioestratigrafía and I Congreso Latinoamericano de Paleontología, 2–6 abril 1978. Asociación Paleontológica Argentina, Buenos Aires, Argentina, pp 237–256

Stuiver M, Braziunas TF (1993) Modeling atmospheric ^{14}C influences and ^{14}C ages of marine samples to 10,000 BC. Radiocarb 35:137–189. https://doi.org/10.1017/S0033822200013874

Vandenberghe D (2004) Investigation of the optically stimulated luminescence dating method for application to young geological sediments. Unpublished doctoral thesis (PhD), University of Ghent, Ghent, Belgium

Villwock JA, Tomazelli LJ, Loss EL, Dehnhardt EA, Horn Filho NO, Bachi FA, Dehnardt BA (1986) Geology of the Rio Grande do Sul Coastal Province. Quat. S. Am. Antarct. Penins. 4:79–97

Geoarchaeology and Historical Archaeology in Pelourinho, Salvador, Brazil: Settlement, Landscape, and Hypotheses

Julio Cezar Rubin de Rubin, Rosiclér Theodoro da Silva, Rosana Najjar, Catarina Menezes Ferreira and Gabriele Viega Garcia

Abstract The geoarchaeological and historical archaeological research conducted under the Pelourinho Project established new viewpoints and discoveries on the urbanization of Salvador's Historic Center, with a focus on its constructed technogenic deposits. Since these deposits suggest intentionality, issues such as city growth, space occupation, constructive processes, the material culture used, and its origin provided a unique context for investigating the cultural and environmental procedures used in the seventeenth and eighteenth centuries, including the intentional reuse of construction waste. The identification of two types of technogenic deposits formed by the presence and intensity of a number of material culture elements was significant.

Keyword Constructed technogenic deposits · Urban occupation · Landfill

Introduction

The present article is the result of the Pelourinho Archaeological Research Project (Najjar and Silva 2006), carried out between March 2006 and December 2009, in order to diagnose and study the area of the seventh stage of the Salvador Historic Center Recovery Project, in Bahia state, Northeastern Brazil. The research project was sponsored by the Historical Heritage and National Artistic Institute (IPHAN). The project was funded by IPHAN (Department of Material Heritage) and the Super-

J. C. R. de Rubin (✉) · R. T. da Silva
Pontifícia Universidade Católica de Goiás, Av. Universitária 1440, Goiânia, GO, Brazil
e-mail: rubin@pucgoias.edu.br

R. Najjar
Instituto de Patrimônio Histórico e Artístico Nacional, Av. Rio Branco 46, Rio de Janeiro, RJ, Brazil

C. M. Ferreira
Universidade Federal de Sergipe, Av. Marechal Rondon s/n., Rosa Elze, São Cristóvão, SE, Brazil

G. V. Garcia
Instituto Homem Brasileiro, Rua dos Coqueiros 19, Jardim das Palmeiras, Cuiabá, MT, Brazil

H. Inda Ferrero and F. García Rodríguez (eds.), *Advances in Coastal Geoarchaeology in Latin America*, The Latin American Studies Book Series,
https://doi.org/10.1007/978-3-030-17828-4_4

intendency of the city of Bahia and especially the MONUMENTA/IPHAN program of the Brazilian Ministry of Culture, which promotes the recovery and preservation of the cultural heritage through economic and social development.

The city of Salvador, the first capital of Brazil, was designated a World Heritage site in 1984 by UNESCO and a Cultural Heritage of Humanity site in 1985. Pelourinho, first colonized in the sixteenth century, is the heart of Salvador's Historic Center. As such, its soil contains valuable archaeological remnants of its successive occupations. In this respect, IPHAN, the main Brazilian cultural heritage preservation entity, enforces the laws governing archaeological heritage, always aimed at producing knowledge that furthers our national identity. The archaeological heritage is protected by the Federal Law 3,924/ 61, Federal Decree n.25/ 37, and Article n.216 of the 1988 Federal Constitution.

The Pelourinho Archaeological Project has been a daunting challenge since its inception, given that it was not merely intended to monitor excavations. The aim was to survey the area of the seventh stage of the Salvador Historic Center Recovery Project (Najjar and Silva 2006) from the archaeological standpoint to better understand the settlement of what is today as the Historic Center. Interdisciplinarity was a basic characteristic of the project and the present article is one of the most striking examples of the effective collaboration of geoarchaeology in order to understand the history of that part of Salvador.

The polygon of the research area covered 11 blocks, with approximately 52,300.00 m^2, and included degraded buildings. The polygon is bordered to the south by the Monte Alverne Street, including the São Francisco Covent, to the north by the Treasury buildings 1 and 2 and the Igreja da Ajuda (Ajuda Church), to the east by part of blocks 30S and 31S on São Francisco Street, and to the west by José Gonçalves Street. A total of 63 buildings were surveyed in all 11 blocks, providing a broad perspective of the different contexts. A number of research methods were used: we conducted a systematic survey (in addition to prospection) in 14 buildings, broad prospection in 29, and detailed registration in 20 (Najjar and Silva 2004).

It is important to underscore that, up to the start of the project, no archaeological research had involved a detailed study of the urban settlement of the Upper City (Cidade Alta) of Salvador, meaning there was no information available to help elucidate these questions. As a result, and with a view to better understanding the study area from the archaeological standpoint, theoretical and methodological procedures were used to establish the questions to be answered and assess the stages of spatial occupation of the Historic Center of Salvador.

Based on the assumption that archaeology is the science of material culture and that architectural objects are also material culture, the theoretic approach adopted was post-processual archaeology (Hodder 1995), which, in turn, assumes that

> (…) objects are not only the result of adaptation, but elements with multiple meanings used by individuals in a society to symbolize their relationships. In this respect, it does not matter if there are dozens or hundreds of objects [or buildings]; it is not the quantity that is important, but the quality. Similarly, (…) it is not society that is at stake, but the actions of each individual in interacting in this society, accepting or resisting its social rules (Najjar 2005b, pp. 15–16).

Based on this perspective, each of the buildings surveyed and the urban structure itself were the research objects, understood as the materialization of several aspects of Salvador's society since its foundation in the sixteenth century. This viewpoint potentially allows us to reveal the existing relations between a number of historical figures, oppressor, and oppressed alike. In this respect, our goal was to use the results of excavations to give a voice to the social groups that lived there. The project also adhered to the procedures established in the IPHAN/Monumenta Historical Archaeology Manual (Najjar 2005b) in an effort to ensure better archaeology research quality and productivity within architectural restoration projects.

The architectural survey started with the analysis of buildings as a minimum observation unit, aimed at comprehensive understanding, where the urban setting was the primary, but not the only focus. The objective was to understand how the area was settled, establishing the stages and respective modes of occupation. Geoarchaeology was an essential analysis tool. In summary, the focus was to study the urbanization of the area over time and observe the presence of individuals that built the city.

Based on the methodological–theoretical approach, historical survey, and the first months of the study, a number of questions and assumptions were raised as the basis for developing the project, including four assumptions that are directly related with geoarchaeology:

1. The importance of viewing the constructed environment as a source of data that allows to interpret the profile of individuals who made up the society in different settlement periods in Salvador, architecture as temporal marker;
2. The need to discover how the growth of Salvador occurred, how urban organization reached the point it is at today, and how the space was occupied;
3. The need to observe the constructive systems used, the differences and/or similarities between the types of techniques and materials used and the settlement context given the topography of the area (settlement patterns). Correlate the constructive systems with the terrain on which they were built based on the diversity of the building and artifact remnants found;
4. The obligation to consider the topographic context of the area under study. A number of overlapping landfills consisting mainly of trash (material culture) were observed during excavations. As such, it is necessary to know the temporal and spatial logic of these landfills and how relevant they were to the occupation of the area.

The present study shows the importance of geoarchaeology in the Pelourinho Archaeology Project, producing unprecedented and exciting data, allowing new viewpoints and discoveries about the urban settlement of the Historic Center of Salvador. This article is an extension of the chapter written by Rubin et al. (2010) for the book entitled "Archaeology in Pelourinho/Organized by Najjar" (2010).

The urban occupation of Salvador was concentrated on the valley floors, especially the Tripas River and its tributaries. Some of these areas were filled with the remains of demolitions, fragments of pottery, glass, metals, bones and shells, trash discarded by the inhabitants of Salvador, serving as the primary basis for leveling the terrain for construction projects. These components were also used to fill between the foundation blocks of buildings. Historic city maps from the seventeenth and eighteenth century show the first urban occupation on hillsides and interfluves, and later on the valley floors. Sampaio (1949) refers to the geography of the area where the Historic Center of Salvador lies today, particularly the trench located between Rua do Tijolo (Rua means 'street' in English, that is, Rua do Tijolo translates as Tijolo Street) and Ladeira da Praça (Ladeira da Praça translates as Hillside Square since ladeira means 'hillside' and praça means square), where this study began. The author cites a letter of Father Manuel da Nóbrega from the early days of the city, which mentions the intention to extend it to the area around the trench.

> The primitive palisade did not extend beyond this; since the entire city of Thomé de Sousa lay between the start of Rua da Misericordia, in Praça do Palácio, and Largo do Theatro (Largo means plaza in English, that is, Largo do Theatro translates as Theatro Plaza), where the terrain begins to descend towards the south. At the seaside, the land drops rapidly, nearly cliff-like; inland, the slopes are less steep, and descend to the brushland, where the stream, whose indigenous name has been forgotten, flows northward and encircles the source, draining into Rio Vermelho (rio means river). In this small environment, four doors opened, flanked by improvised bulwarks: the one to the north, at the start of Rua da Misericordia, where there was a natural dip in the land, serving as a trench, which was later filled with trash at Rua do Tijollo; (…) (Sampaio 1949, p. 184),

> (…) at Rua do Tijollo, a very notable depression, which caused the lands to the north to take on the aspect of a steep slope or hill, to which the natural vegetation gave an even more pronounced relief. It is this steep slope that Father Nobrega is referring to in his letter to Father Simão Rodrigues, when he mentions the location for the College, saying: "…it seemed better to all of us that the city should be extended on the steep hill just beyond the fence…" (Sampaio 1949, p. 194)

> (…) in view of the elevated riverbank and the deep grotto, today concealed by the buildings between Rua do Tijolo and Ladeira da Praça. The latter street did not exist yet, just a deep trench which descended to the first bulwark, defending this north face of the walled city, and extending to near what is now known as Praça de Guadelupe (Sampaio 1949, p. 195).

The area discussed here is one of the blocks surveyed, the former Freguesia da Sé (Nascimento 1986, p. 30). The valleys in Brazilian urban areas were settled primarily by low-income populations, since they were unable to acquire costly urbanized land. In the case of Freguesia da Sé, some structures, dimensions, and construction techniques of many homes built on landfills indicate the residents had considerable purchasing power. This may be related to "urban stress" caused by the nonavailability of land on the slopes or interfluves, financial issues, or a new settlement trend.

Research shows that from the mid-nineteenth century, the social profile of this neighborhood began to change, and large homes intended for wealthy families were subdivided to house lower income residents (Nascimento 1986, p. 68). During that century, the area was settled by a mixture of Salvador society, initially and in lower numbers, by white slave owners that were public servants, merchants, the bourgeoisie, or aristocracy, who lived in stately homes or mansions. Later, it was occupied primarily by blacks and mulattos, who may have been artisans or street peddlers living in rammed earth houses and afterward in subdivided mansions (Mattoso 1992, pp. 440–411).

Before it was inhabited, this area was used for waste disposal, and at the same time adapted for building construction. Settlement occurred simultaneously to urban transformations resulting from discourse opposing the disposal of trash in public places and waterways such as the Tripas River, a common practice at the time (Mattoso 1992, p. 442). This did not occur only in Salvador. According to Avila (2010), in 1837 in Porto Alegre, Rio Grande do Sul state in Southern Brazil, the Police Code of Conduct established trash disposal sites, which included riverbanks.

In Salvador, this behavior was still common in the nineteenth century, but began to be addressed mid-century by public policies that sought to improve urban sanitation through codes of conduct. In 1856, the inhabitants of Salvador were encouraged to clean areas considered disease disseminators, a practice that became more pronounced when a deadly 2-year cholera epidemic broke out in 1855 (Mattoso 1992, pp. 442–443). The outbreaks were confined to riverbank and swampy areas of the city (David 1996, pp. 50–63). At the time, it was believed that disease was caused by miasmas and putrefied odors that spread over non-constructed areas commonly used for waste disposal. The solutions adopted included channeling rivers and building landfills, which improved building conditions and eliminated trash disposal sites. The effect was an increase in living areas in a number of neighborhoods, with the construction of large landfills.

The Brazilian archaeological literature reports that landfills were also a means to prepare the land for occupation in other cities and states (Orser 1992). Andrade Lima et al. (1989) found landfills at the Casa dos Pilões archaeological site in the city of Rio de Janeiro; the stratigraphy exhibited by Mentz Ribeiro et al. (2004) for Sobrado dos Azulejos, in the city of Rio Grande, Rio Grande do Sul state also describes the presence of landfills. Symanski and Osório (1996, p. 45) reported that between 1842 and 1870 in Porto Alegre's Praça XV (XV Square), "waste produced by different activities were discharged onto the alluvial plain of the Guaíba River." With respect to the city's public market (Mercado Público), the authors raise the possibility that during building site preparation, "trash in adjacent areas was removed and incorporated into this plot of land." Tocchetto and Capelletti (1995), Landa (1996) also describe landfills in the same city.

These landfills can be classified as technogenic deposits originating from human activity, defined in studies conducted by researchers such as Chemecov (1983), Sergeev (1980, 1986), Kowalski (1984), and Ter-Stepanian (1988). In Brazil, the most important studies were performed by Oliveira (1990, 1994), Rohde (1996), and Peloggia (1998a, b, 1999). Oliveira (1994) classifies technogenic deposits as constructed (landfills, dumps, etc.), induced (resulting from disordered land settlement) or modified (fertilizer application, etc.). Peloggia (1998a) and Nolasco (2002) also presented classifications based on Oliveira (1994).

Cunha (2002), Rubin (2003, 2004), and Rubin et al. (2007, 2008) characterized the technogenic deposits in the urban area of Goiânia, Goiás state, associating them with the main watercourses. The authors provide growth and distribution estimates and correlate them with the founding of the city in 1937. The effect of human activity on nature, modifying its geography, has been well documented in Brazil since its discovery, primarily in cities such as São Vicente and São Paulo (both in São

Paulo state), Rio de Janeiro, Porto Seguro (Bahia state), São Luís (Maranhão state), Salvador (Bahia state), Recife (Pernambuco state), and Ouro Preto and Diamantina (both in Minas Gerais state).

The archaeological and geological excavations of a number of residences and in the courtyard of the fire brigade headquarters located in the study area found different thicknesses of constructed technogenic deposits (CTDs) along the valley floors (up to 12 m), in an area of approximately 52,000 m^2 (polygon of the 7th stage of the Salvador Historic Center Recovery Project). The foundation blocks of several walls measuring up to 6 m high lie on these deposits, as well as some residential columns. The area of the project stands out for its magnitude and especially the number of exposed CTDs, which favored its characterization as limited exposure. The classification of CTDs reflects the situation observed in houses in block 31 and the fire station courtyard, demonstrating their importance and deliberate use in the area.

Materials and Methods

Wells and trenches were dug to identify, describe, and characterize the stratigraphy of the sections. The variables used were particle size and morphometry (roundness and sphericity), shape, color, thickness, and sedimentary structures of the layers, as well as the presence of cultural remnants. To that end, mechanical surveys were conducted at a number of points, but only the one carried out in house 24 was used.

With respect to the excavations performed in the fire station courtyard, it is part of a monitoring initiative of a linear undertaking (trench opening), approximately 170 m long, with a maximum depth of 3.20 m and width between 1.60 and 0.80 m, which extended beyond the area of the building and block 31. For methodological reasons, it was decided to divide the trench into three sectors:

- Sector 1—Inner courtyard of the fire station (Segments A, B, C);
- Sector 2—Grounds of buildings 24, 33, 35, 37, and 39, block 31 up to Rua São Francisco;
- Sector 3—Rua São Francisco and block 28.

Each sector was divided into segments, then into 2 m sections, between manholes. This procedure favored the location, mapping, and recording of archaeological remains.

The cultural remnants were cleaned, catalogued, and classified in the laboratory, and the sediments underwent particle size and mineralogical analysis. Two types of CTDs were established for the area in order to determine the significance of the deposits and systematize the approach:

CTD 2—normally found between the surface and 1.20 m, exhibiting different foundation structures, sewer drainpipes, floors and subfloors, and burial sites with a large amount of cultural remnants such as debris, pottery, glass, metals, bones, and shells, in fine to coarse sand.

CTD 1—located in layers deeper than 1.20 m, characterized by the presence of demolition debris, and less pottery, glass, metals, bones, and shells than in CTD 2, also in fine to coarse sand.

The basic difference between the types of deposits is the amount of waste. Periodic observations revealed that some of the 0.30 × 0.30 m sections of CTD1 contained 30–50% waste, while in CTD2 it reached 80%.

Satellite images, geological maps, geomorphology, pedology, historical city maps, and the topographical map of the study area were interpreted. Moreover, field and laboratory data were interpreted, which were the basis used to formulate hypotheses regarding settlement of the area.

Results

Block 31 Houses

These are located in one of the last blocks occupied due to its rugged topography. According to the historical and archaeological records, it became part of the urban structure in the mid-eighteenth century (Najjar and Silva 2006, 2007).

A number of continuous plots were uneven (up to 2 m in height) in the gradient of the CTDs, which may be associated with the topography, thickness of the deposit, or the level in relation to the rock/soil or fluvial dynamics. In this case, the landfill on a marginal dike, invariably found on a higher gradient than a floodplain or canal, depending on the thickness of the landfill, will also have a higher gradient. Figures 1, 2, 3, 4, 5, and 6 provide details of the deposits.

The nomenclature used in block 31 establishes that the floor is the ground or pavement formed by rock fragments (predominantly pebbles and blocks), cement, and sand; subfloor is the base of the floor, also consisting of sand, cement, and rock fragments (blocks); foundation block is the structure that serves as the base for walls, similar to subfloor, but with a number of boulders.

Figure 6 shows the deposit identified in house 39, at Rua de Setembro 28, block 31, the valley floor of a Tripas River tributary, where it is nine meters thick (determined by percussion probing), and consists of the remnants of crockery, pottery, bones, tiles, shells, etc. (CTD 2). A slight unconformity (this term has been adapted from the geological sciences) which means "a buried erosional or non-depositional surface (…), which can correspond to a time interval (…) necessary for deposition of a simple to very long layer" (Suguio 1998, p. 236) separates different units (Fig. 3), which is believed to be the result of the landfill construction process—a flow of waste, arranged according to the position in which it was discarded. At certain points, the deposition is subhorizontal and in others the dip angle is approximately 45°, which may indicate the direction of the construction. These deposits were found in manual excavations and percussion probing.

Fig. 1 Foundation block filled by CTD2. In the lateral trench, part of the foundation block of the wall. House 39, Rua de Setembro 28, block 31

On the other hand, Figs. 2 and 6 highlight the relation between CTD1, CTD2, foundation block, subfloor, and floor, a rare opportunity to observe the elements used in the appropriation and construction process of the landscape in the area.

The Fire Station Building

The fire brigade headquarters is located at block 31, at the corner of Ladeira da Praça and Rua J. J. Seabra. The building, which exhibits modern architecture, was built in the first half of the twentieth century. In November 2008, a sewer system was implemented for homes constructed in the seventh stage of the Salvador Historic Center Recovery Project. Complying with a request from UEP/Monumenta/Conder, the Pelourinho Project archaeological team supervised the undertaking. UEP refers to Project Execution Unit;

Monumenta is a strategic Ministry of Culture program that seeks to recover and preserve cultural heritage through economic and social development (Najjar 2010, p. 11).

Conder is an urban development company of the state of Bahia.

Fig. 2 Trench from Fig. 1 with CTD, foundation block, subfloor, and floor. Note the circle around the fragment of crockery, also highlighted in Fig. 3. Depth = 1.5 m

Excavations also made it possible to determine how CTDs were structured. However, over the years, they have undergone numerous interventions by construction projects that occurred after deposition. Below the asphalt layers is the cobblestone floor, fine sand, earthen floor, and a structure made to install a sewage drainpipe that passes through two layers of CTDs. Despite the interventions, it was possible to characterize the well-preserved conditions, which enabled a reliable interpretation of CTD depositions.

In this study, only consolidated research of sector 1 is discussed. This involves the inner courtyard of the fire brigade building, segments A, B, and C, focusing on a number of structures with relieving arch foundations. Two of the structures found under the current floor were noteworthy due to their location, robustness, type of construction material, and mainly for lying over CTDs, more than three meters below the asphalt. They are aligned parallel to one another five meters apart. They are likely remnants of the old Capela de Nossa Senhora de Guadalupe (Our Lady of Guadeloupe Chapel). The Archaeological Research Project considers both CTDs and architectural structures as "super-artifacts" produced by man at a particular time and space.

Fig. 3 Typical CTD2 with predominance of demolition debris, pottery, bones, and shells. Highlight of the crockery (yellow arrow) and part of the dental arch (red arrow) shown in Figs. 4 and 5. The dashed lines correspond to the unconformities. Depth = 1.5 m

Fig. 4 CTD2 from Fig. 2 showing pottery fragments, part of a dental arch and sediments

Fig. 5 CTD from Fig. 4, with a large amount of pottery fragments. The arrows indicate the predominant directions of the dip angles of the tailings

Fig. 6 Arrangement of CTD1, CTD2, foundation block, subfloor, and floor. House 41, Rua de Setembro 28, block 31. Depth = 1.8 m

A total of approximately 144 m^2 of sediments were turned over. During the excavation, 505 fragments of cultural remnants were collected (pottery, glass, metal, etc.). The emergence of burial sites required systematic prospections to record the recovery of 12 intact skeletons found at the axis of the sewage network. Some of the architectural structures were cut and cleaned for identification and registration purposes.

A number of the cultural remnants such as earthenware and coins recovered and previously analyzed indicate that these structures were built in the second half of the eighteenth century, which coincides with the information found in the historical settlement records for the area. This is the same period that the Capela de Nossa Senhora de Guadalupe was constructed (Najjar 2005a).

The primary documents found by project historians confirm that the Capela de Nossa Senhora de Guadalupe was built in 1778 to comply with a request made by the mulatto sisterhood of Igreja da Sé (Igreja means Church). APEB (the Public Archive of Bahia State), Colonial Section, Government of the Province, Religion, produced a book of designated historical heritage (Livro de Tombo) of the assets of all third orders, brotherhoods, and sisterhoods of Salvador, 1851–1853, recorded as nº 5266. Also according to historical information, it was demolished in 1857. This may have occurred due to physical or political factors of that period such as the poor state of conservation of the building, perhaps because it is lying on CTDs; its geographic location on the banks of the Tripas River; serious sanitary problems that the city experienced at the time; and primarily, due to the channeling of the Tripas River, as stated in the document below:

> (…) in compliance with that determined by your excellency in the bulletin of the 19th of the current month, with which you honored me, I have summoned the Sisterhood of Our Lady of Guadeloupe on Sunday the 23rd of this month, and a commission was appointed to better understand the proposal made by the winning bidder of the "valla" undertaking, regarding the materials belonging to the chapel as well as its final demolition: which will occur in the next eight days. May God keep your Excellency for many years. Bahia 25 of August of 1857. (Source: APEB, Colonial Section, Government of the Province, Religion, Sisterhood of the Capital, 1847–1889, Record no. 5249.)

The burial sites identified in the trench between 0.90 and 1.20 m deep and covered by CTDs were well preserved, some surrounded by a layer of lime, with some intact skeletons, and others fragmented. The position of the 12 individuals was compatible with a Christian burial. The arrangement of their positions denotes primary burials. Together with these skeletons were cultural remnants such as necklace beads, hair clips, buttons, etc., which help identify their social class.

When investigating historical questions that emphasize the constructions and lifestyle in a populous area such as the center of Salvador between the eighteenth and twentieth centuries, it has been observed that the buildings in block 31 had to cope with the channeled Tripas River in the nineteenth century. This watercourse served as the dumping ground for all types of putrid waste from pigsties, slaughterhouses, fish markets, tanneries, and sewer pipes. Moreover, all the hospitals located near its banks discarded their waste in the river (David 1996, p. 55). Beyond the channeling of this river and the unsanitary issues raised, the archaeology and geoarchaeology of this study contribute to the history of the city by drawing attention to the landfills and their compositions that are frequently cited by different historians, but have never before been studied.

The Landfill—CTD

As previously mentioned, geological information shows that the area where blocks 28 and 31 are located have CTDs up to 9 m deep, indicating a rugged topography and rossio-type terrain, which existed in the first centuries of the city's occupation. In line with Portuguese tradition, the city of Salvador was created with a "termo" and a "rossio", the former designating all the territory under municipal authority. The rossio was part of the termo, located near the urban space, and was used by all the inhabitants as a grazing area and to supply firewood, the main fuel. It was also open to urban expansion. This answers one of the questions raised by the project: why did an area so close to the center take so long to be densely settled? It also shows that CTD was a prerequisite for implementing new blocks in the urban structure and consequently new housing. Characterizing CTDs made it possible to investigate aspects of their deposition (if deposits overlapped and if they were compacted or not) and functionality (if they were aimed at supporting the foundations of the mansions or only to fill the spaces between the foundation blocks of the buildings or serve as the subfloor).

The CTDs identified in the fire brigade building along the trench (sector 1) are similar to those in block 31 houses, also subdivided into CTD 1 and 2, according to the position they occupy in the stratigraphic package, indicating a pattern for the region.

The two types of CTDs are also differentiated by the specific type of material remnants found. In this separation, the CTD1 contained no English pottery, whose production began in the late eighteenth century and was exported to Brazil throughout the nineteenth century. This suggests that CTD1 is older than CTD2 or, if contemporary, has different origins that need to be investigated.

This material culture is associated with the different moments of settlement in the area, favored by the fact that a number of sites were less disturbed, enabling detailed characterization and classification. In general, the following stratigraphic layers were identified in the area: asphalt or thick concrete layer; cobblestone floor and/or subfloor made of carved stone; fine sand and/or CTD2 base; washed beach sand lens and malacological material; earthen floors (clay and lime mortar); CTD2 and CTD1.

Final Considerations

Miyazaki (2014) underscores the importance of technogenic deposits in the processes of appropriating and settling urban areas, which corroborates the findings presented here. Rundic et al. (2012) showed a correlation between archaeological sites and technogenic deposits in the Danube River in Belgrade, Serbia. Peloggia (2015) describes the use of alluvial technogenic deposits in an investigation of the settlement of the Una River Basin in São Paulo state in the nineteenth century, when the Atlantic Forest was replaced by coffee plantations, similar to the discussions of Rubin et al. (2008).

The constructed technogenic deposits identified in the Pelourinho Project, associated with the historical archaeology, provide a rare opportunity to observe a number of cultural and environmental procedures used in the seventeenth and eighteenth centuries, such as the intentional reuse of building waste materials, since this type of deposit indicates intentionality, as well as landscape-altering procedures aimed at urban settlement, similar to what is currently occurring in large- and medium-sized Brazilian cities.

In this respect, the present investigation contributes to archaeological and geoarchaeological landscape research, as in Criado-Boado (1999), in that it considers landscape as the result of man's impact on the environment. The results obtained enable a basic understanding of the appropriation and construction of the landscape in a given time period. In addition to documenting the physical aspects and their traits, studies on the types of landfills presented here open a dialogue with the people who built these landfills and with the discourses and practices of life at that time.

Another question raised is the volume of waste used. Considering the average waste thickness of four meters in an area of 52,000 m^2 of the polygon of the seventh stage of civil works, a value of approximately 208,000 m^3 was obtained. Historical documents indicate that the city was built around 1625 after the Portuguese and Spanish reconquest, which may be the origin of a considerable amount of the waste used.

The CTDs may also be related to the previously mentioned poor state of conservation of Capela de Nossa Senhora de Guadalupe, constructed over alluvial deposits from the Tripas River. This resulted in overlapping deposits, perhaps during the restructuring of an area with geotechnical conditions unfavorable for building, aggravated by the water table near the surface and the constant overflowing of the Tripas River. Moreover, the CTDs in this area are covering burial sites.

In addition to the aforementioned aspects, the planes of unconformity and the direction of the dip angle of the deposits make it possible to establish hypotheses about the building process. Figure 3 raises the hypothesis that deposition occurred predominantly from left to right in the photographed area. A general survey of the profiles may establish the processes adopted in the landfills of the area, a significant fact for historical archaeology, geoarchaeology, and environmental geology.

Acknowledgements The authors would like to thank the project team, especially the archaeologists Cláudio César de Souza e Silva and Samuel Lira Gordenstein. Thanks also to the archaeologist Joanne Ester Ribeiro Freitas and the architect Adriano Costa Bandeira for the preparation of the figures.

References

Andrade Lima T, Fonseca MPR, Sampaio AC de O, Fenzl-Nepomuceno A, Martins AHD (1989) A tralha doméstica em meados do século XIX: reflexos da emergência da pequena burguesia do Rio de Janeiro. Dédalo 1:205–230

Chemecov YF (1983) Tecnogenics deposits. In: 11st INQUA Congress, Moscow 2008, vol 62

Criado-Boado F (1999) Del terreno al espacio: planteamientos y perspectivas para la arqueologia del paisaje. Arqueologia del Paisaje. In: CAPA 6, Grupo de Investigación em Arqueología del Paisaje, Universidad de Santiago de Compostela, pp 1–58

Cunha BCC (2002) Impactos sócio-ambientais decorrentes da ocupação da planície de inundação do ribeirão Anicuns: o caso da Vila Roriz. Goiânia. Dissertação, Universidade Federal de Goiás

David OR (1996) O inimigo invisível: epidemia na Bahia no Século XIX. EDUFBA/Sarah Letras, Salvador

de Avila VF (2010) Histórias e práticas cotidianas sobre o saneamento: desdobramentos na Porto Alegre do Século XIX (1850–1900). Dissertação, Pontifícia Universidade Católica de Rio Grande do Sul

Hodder I (1995) Theory and practice in archaeology. Routledge, London, New York

Kowalski W (1984) History of change of geological environment under the influence of the activity of mankind. VNU Science Press, Ultrech, vol 17, pp 51–67

Landa BS (1996) Acompanhamento arqueológico no Mercado Público Central de Porto Alegre. Revista do CEPA 20(23):77–104

Mattoso KM de Q (1992) Bahia Século XIX: uma província no Império. Editora Nova Fronteira, Rio de Janeiro

Mentz Ribeiro PA, Pestana MB, Penha MAP (2004) Escavações arqueológicas no Sobrado dos Azulejos, Rio Grande, RS, Brasil. In: Biblos, vol 16, pp 201–228

Miyazaki LCP (2014) Depósitos tecnogênicos: uma nova perspectiva de leitura geográfica. Quat Environ Geosci 5(2):53–66

Najjar R (2005a) Construtores de Igrejas: um estudo arqueológico da presença da Companhia de Jesus do litoral brasileiro. Universidade de São Paulo, Tese

Najjar R (2005b) Manual de arqueologia histórica em projetos de restauração. IPHAN, Brasília

Najjar R (2010) Arqueologia no Pelourinho. IPHAN/Programa Monumenta, Brasília, DF

Najjar R, Silva RCP (2004) Projeto de prospecções arqueológicas na área da 7ª Etapa do Pelourinho. IPHAN/ MONUMENTA

Najjar R, Silva RCP (2006) Relatório parcial do projeto de prospecções arqueológicas na área da 7ª Etapa do Projeto Pelourinho. IPHAN/MONUMENTA

Najjar R, Silva RCP (2007) Relatório de pesquisa arqueológica da 7ª Etapa do Projeto Pelourinho Salvador/BA. IPHAN/MONUMENTA

Nascimento AAV (1986) Dez freguesias da cidade do Salvador: aspectos sociais e urbanos do século XIX. Fundação Cultural do Estado da Bahia, Salvador

Nolasco MC (2002) Registros geológicos gerados pelo garimpo. Lavras Diamantinas – BA. Tese, Universidade Federal do Rio Grande do Sul

Oliveira AMS (1990) Depósitos tecnogênicos associados a erosão atual. In: 6 Congresso Brasileiro de Geologia de Engenharia, Salvador, Out 1990, pp 411–415

Oliveira AMS (1994) Depósitos tecnogênicos e assoreamento de reservatórios: exemplo do reservatório de Capivari, Rio Paranapanema, SP/PR. São Paulo. Tese, Universidade de São Paulo

Orser CE (1992) Introdução à arqueologia histórica. Oficina de Livros, Belo Horizonte
Peloggia AUG (1998a) O Homem e o ambiente geológico: geologia, sociedade e ocupação urbana no município de São Paulo. Xamãn, São Paulo
Peloggia AUG (1998b) A magnitude e a frequência da ação humana representam uma ruptura na processualidade geológica na superfície terrestre? Geosul 14(27):54–60
Peloggia AUG (1999) Sobre a classificação, enquadramento estratigráfico e cartografação dos solos e depósitos tecnogênicos. In: Peloggia AU (ed) Estudos de Geotécnica e Geologia Urbana (I). Manual Técnico 3 GT – GEOTEC. Superintendência da Habitação e Desenvolvimento Urbano, Prefeitura do Município de São Paulo
Peloggia AUG (2015) O significado arqueológico dos depósitos tecnogênicos aluviais da bacia do rio Una (Taubaté, SP). Rev Arqueologia Pública 9(11):207–219
Rohde GM (1996) Epistemologia ambiental. EDIPUCRS, Porto Alegre
Rubin JCR (2003). Sedimentação quaternária, contexto paleoambiental e interação antrópica nos depósitos aluviais do alto rio Meia Ponte, Goiás. Tese, Universidade Estadual Paulista
Rubin JCR (2004) A relação do ser humano e o rio Meia Ponte em Goiânia (GO). Fragmentos de Cultura 14(7):1275–1285
Rubin JCR, Brito GS, Oliveira AMS, Saad AR (2007) Depósitos tecnogênicos associados a um segmento da planície aluvial do ribeirão Anicuns, Goiânia, GO. In: XI Congresso da Associação Brasileira de Estudos do Quaternário, Belém, vol 45
Rubin JCR, Oliveira AMS, Saad AR, Brito GS (2008) Amostragem dos depósitos tecnogênicos associados ao rio Meia Ponte na área urbana de Goiânia – GO. Revista Brasileira de Geomorfologia 9(2):3–14
Rubin JCR, Silva RT, Garcia GV (2010) A geoarqueologia no Projeto Pelourinho. In: Najjar R (org.) Arqueologia no Pelourinho. IPHAN/Programa Monumenta, Brasília, DF, pp 104–115
Rundic L, Knezevic S, Kuzmic V, Kuzmic P (2012) New dates on the geology of the archaeological site at Vinca (Belgrade, Serbia). Annales Géologiques de la Peninsule Balkanique 73:21–30
Sampaio T (1949) História da fundação da cidade do Salvador. (Obra póstuma, publicada por indicação de AnísioTeixeira). Tipografia Benedita Ltda/Bahia, Salvador
Sergeev YM (1980) Engineering geology and protection of the environment. Bull Int Assoc Eng Geol 22:75–78
Sergeev YM (1986) Theoretical fundamentals and problems of engineering geology. In: Proceeding of International geological congress, 27, Moscow. VNU Science Press, Ultrech, vol 17, pp 35–49
Suguio K (1998) Dicionário de geologia sedimentar e áreas afins. Bertrand Brasil, Rio de Janeiro
Symanski LCP, Osório SR (1996) Artefatos reciclados em sítios arqueológicos de Porto Alegre. Rev de Arqueologia 9:43–54
Ter-Stepanian G (1988) Beginning of the tecnogene. Bull Int Assoc Eng Geol 38:133–142
Tocchetto F, Capelletti A (1995) A intervenção arqueológica em Porto Alegre e o exemplo de dois sítios históricos na área central da cidade. Manuscript, Porto Alegre

The Geoarchaeology of Intertidal Shipwreck Sites: Case-Studies from Highly Dynamic Settings in Southern Brazil, Argentina and Uruguay

Rodrigo Torres, Dolores Elkin, Roberto Kokot, Daniela De Hoz, Eduardo Keldjian and Matías Dourteau

Abstract Shipwreck sites often occur in highly dynamic intertidal environments, posing major challenges in terms of preservation and coherence of the archaeological record. These sites typically consist of widely scattered materials located either in the intertidal zone or above the high waterline. Exposure to winds, waves, surges, tides, currents, and sediment transportation occurs continuously across space and over time, and within various significant spatial-temporal scales, resulting in natural cycles of flood, burial and exposure of stranded wrecks. Ultimately, however, the survival of shipwrecks in the archaeological record in these environments will depend on an array of interactions with natural and cultural post-depositional processes. The following chapter presents the current knowledge regarding the archaeology of intertidal shipwreck sites in South America, based on case studies from highly dynamic settings in Southern Brazil, Argentina and Uruguay. It aims at highlighting regional themes and defining the gaps which need to be filled towards a more systematic macro-regional approach.

R. Torres (✉) · E. Keldjian · M. Dourteau
Programa de Arqueología Marítima y Subacuática, Centro de Investigaciones del Patrimonio Costero, Universidad de la República, Tacuarembó entre Avenida Artigas y Aparicio Saravia, 20000 Maldonado, Uruguay
e-mail: rodrigo.torres@cure.edu.uy

D. Elkin
CONICET and Programa de Arqueología Subacuática, Instituto Nacional de Antropología y Pensamiento Latinoamericano, 3 de Febrero 1378, 1426 Ciudad de Buenos Aires, Argentina

R. Kokot
Facultad de Filosofía y Letras, Universidad de Buenos Aires, Puan 480, C1420 Ciudad de Buenos Aires, Argentina

D. De Hoz
Programa de Arqueología Subacuática, Instituto Nacional de Antropología y Pensamiento Latinoamericano, 3 de Febrero 1378, 1426 Ciudad de Buenos Aires, Argentina

H. Inda Ferrero and F. García Rodríguez (eds.), *Advances in Coastal Geoarchaeology in Latin America*, The Latin American Studies Book Series,
https://doi.org/10.1007/978-3-030-17828-4_5

Keywords Shipwrecks · Site formation processes · Beach environments · Coastal dynamics

Introduction

Every archaeological site poses its own set of challenges for the study and interpretation of the material culture therein. As Harris (1989: xi) noted, the particular characteristics of a depositional setting of the archaeological record, will depend upon the historical, cultural and environmental circumstances in which it was created and transformed. By definition, an archaeological site is a place which contains, in a more or less preserved state, information about past events and activities embedded in artifacts and their spatial relationships (Willey and Phillips 1958: 18; Binford 1964: 431; Burke and Smith 2004: 63; for a nuanced discussion, see Dunnell 1992). It relies upon the archaeologist, however, to define an operational unit which accounts for historical and contextual significance within a given analytical framework and research designs.

Contexts constitute the raw material of archaeological inquires, and refers to the relationships that artifacts and features have to each other and the situation in which they are found. It is the spatial as well as the chronological relationship, established between artifacts on the same site and between different sites within a landscape. At the fundamental level, the context of a find, whether it is an artifact, feature or structure, consists of its surrounding sedimentary matrix, its provenience within it, and its association with other finds (Renfrew and Bahn 2008: 52–54).

Perhaps the principal contribution to mainstream archaeology from site formation processes theory is that the observed archaeological record is not necessarily a direct expression of past human behavior, thus giving rise to inference and interpretation issues (see, for example, Schiffer 1976, 1996). This attention to problems related to inference from the archaeological record becomes even more critical as dealing with archaeological sites in highly dynamic depositional settings, since materials are continuously being added, removed, and rearranged, such as in the case of beach stranded shipwreck sites. It is often assumed in these circumstances that cultural relatedness is too highly variable and does not correlate with spatial proximity, drastically affecting the coherence of the archaeological record. These sites typically consist of widely scattered materials located either in the intertidal zone or above the high waterline. Specialized literature has referred to these sites as 'scattered' or 'scrambled' (Delgado 1984, 1985, 1997; McNinch et al. 2006; O'Shea 2002; Russel 2002; Muckleroy 1975).

In fact, concerns with understanding shipwreck site formation have followed the field of shipwreck archaeology since its origins in the 1960s (Dumas 1962; Bass 1966). It initially developed around issues of defining shipwreck site typologies and discussing the coherence of the archaeological record in relation to the physical characteristics of the underwater environment (Dumas 1962, 1972; Frost 1962; Muckelroy 1977).

Due to the nature of incredibly rich shipwreck sites along the Mediterranean coasts, where these 'microcosms' of ancient people have been investigated since the 1960s, the sub-discipline of nautical archaeology had its beginnings associated by general public with the excavation of well-preserved, near-pristine shipwreck sites (see Bass 1972, 1983). Early publications have somewhat overlooked the significance of shallow and scattered ship remains, due to their presumed lack of coherent information content (Dumas 1962, 1972; Frost 1962).

During these formative stages of underwater archaeology, the idea of pristinely preserved 'time-capsules' pervaded the imagination of archaeologists and the general public, supporting the perception of shipwreck archaeology as inherently linked to the study of unique snapshots of single, dramatic past events frozen in time (see, for example, UNESCO 1972). In fact, these early works were operating largely under what has been called the 'Pompeii premise' (Ascher 1961; Binford 1981). Despite the paramount importance of these works for the development of the shipwreck archaeology, this perspective has somewhat biased the interest of the discipline, relegating the investigation of scattered and less coherent sites to a less important corner of archaeology.

Broader scope studies of the processes affecting the archaeological record first appeared explicitly in the work of processualist and behaviorist archaeologists concerned with the inferential problems of linking the archaeological record to past human behavior, thus inaugurating a new paradigm in the discipline of archaeology (Schiffer 1972, 1976; Binford 1981, 1983). Central to site formation theory is the argument that these processes operate in regular and predictable ways, so the knowledge of the factors involved in shaping an observed archaeological record allow archeologists to recognize and work retrospectively to understand behavior patterns as they existed.

Typically identified with the contribution of Michael Schiffer, site formation theory has established a framework for the consideration of cultural and natural processes transforming material culture within archaeological settings (Schiffer 1972, 1976, 1996; Schiffer and Rathje 1973). Schiffer observed that artifacts, as well as features and sites, transit between systemic contexts, wherein artifacts are part of an ongoing socio-cultural system of manufacture, use, reuse, and discard, and archaeological contexts, where those artifacts culturally deposited or lost are no longer part of an ongoing society. In fact, artifacts can move back and forth within these different contexts during their life histories.

Explicit scientifically-oriented, pattern-searching, statistically-laden frameworks in formation processes for the field of maritime archaeology gained prominence with the contributions of Muckelroy (1975, 1978). Working in underwater sites off the coast of Scotland, Muckelroy made a seminal contribution to the interpretation of shipwreck sites by emphasizing shipwreck site formation as a process and highlighting the role of both cultural and natural processes acting as 'extracting filters' and 'scrambling devices'.

Considered the first attempt to develop and apply explicit middle-range theory for maritime archaeology, Muckelroy's conceptualization of shipwreck site formation processes continues to be influential more than 30 years after its publication

(Adams 2009; Harpster 2009; Gibbs 2006). He generated more inclusive and comprehensive site classifications than his predecessors, identifying five different classes of survival of archaeological material, ranging from well preserved and coherent remains, to widely scattered, disarticulated sites. Moreover, the author developed the first methodological program for the interpretation of scattered shipwreck sites (see Muckelroy 1975, 1977, 1978).

Muckelroy's two stage method discussed in 1975 presented an efficient way to approach scattered wreck remains, relating recording precision and significance of the archaeological material. However, it considers a rather strict definition of site and does not account for contextual relationships when evaluating the site's potential. This has been somewhat overcome in works dealing with broader concepts for site definition, such as of maritime landscape and shipwreck province (Westerdahl 1992; Duncan 2000; O'Shea 2002), where interpretations benefit from an interplay between scales of analysis within regional settings.

Traditionally, most contributions to shipwreck formation processes focus on the physical, biological, chemical and sedimentary factors responsible for shaping the archaeological record (Murphy 1997; Stewart 1999; Wheeler 2002; Quinn et al. 2007; Keith 2016). While working along these lines, some researchers have attempted to formulate generalized and universal-like models for shipwreck disintegration, introducing a predictive quantifiable dimension in the study of wreck sites (McCarthy 1998; Ward and Larcombe 2002; McNinch et al. 2006).

Ward et al. (1999), for instance, have used differential calculus and derivatives to pursue a process-oriented, scale independent, model to explain the disintegration of shipwrecks. In another interesting work, O'Shea (2004) applied Bayesian probabilistic approach to assess and quantify the level of confidence that can be attached to the historical identification of scattered wreck sites in the Great Lakes region.

From the late 1980s, however, the interest in cultural factors behind shipwreck distribution and wreck assemblages has grown in the specialized literature. In an influential work, Donna Souza identified a range of cultural pre-depositional behaviors influencing the formation of the archaeological record, such as risk minimization strategies, as well as depositional and post-depositional factors, including jettisoning, refloating of stranded vessels and salvage choices in Dry Tortugas, Florida (Souza 1998).

O'Shea (2002), based on a regional survey of shore stranded historical shipwrecks (19 and 20th centuries), and drawing upon analytical categories present on the works of Souza (1998), Schiffer (1996) and Clarke (1973), proposed a systematic approach for studying the implications of ship construction, use, loss, deposition, and recovery under the so-called formation theory. O'Shea also adapted theories from related disciplines, such as oceanography, geology, zoology and physics to explore very specific empirical data and examine general patterns concerning pre-depositional (systemic context), depositional (shipwrecking) and post-depositional processes (site alteration) in a regional scale.

Stemming from the Australian school of maritime archaeology, other contributions have highlighted several past and present cultural and political instances shaping the archaeological record (McCarthy 1998; Richards 2002, 2003, 2004). Richards

(2002), for instance, used historical and archaeological data to analyze the remnants of deliberately abandoned watercraft in Australia, noting how discard and demolition practices were intimately connected to economic trends and technological developments throughout the many phases in the life history of a vessel.

Drawing upon the well-developed literature on disaster studies, Martin Gibbs explored characteristics of shipboard behavior, wreck site formation and the 'archaeological signature' of human responses to shipwreck catastrophe as cultural phenomena (Gibbs 2006). Brad Duncan, another Australian maritime archaeologist, proposed an interpretative framework based upon the concept of maritime cultural landscape to study how seascapes are shaped through human responses and perception of risk (Duncan 2000).

Generally speaking, it is understood that the occurrence and survival of shipwrecks in an archaeological record will ultimately depend on the probability of occurrence of accidents, the characteristics of the ship and the circumstances of the accident, and its interaction with natural and cultural post-depositional processes. The discussion in the following section approaches those aspects of the formation processes related to the interaction between shipwrecks and the beach environment.

Beach Geomorphology and Coastal Dynamics

Beach environments are among the most energetic and complex depositional systems in the world's oceans (King 1972). Winds, waves, storm surges, tides, coastal currents, and consequent sediment transportation are the main natural forces affecting the archaeological record in intertidal beach environments. Interaction between these elements occurs continuously across space and over time, and within various significant spatial-temporal scales, resulting in natural cycles of flood, burial and exposure of stranded wrecks (Fig. 1).

Wind fields generated from weather systems off shore act upon the sea surface creating waves, which eventually break upon shore, originating currents, and providing most of the energy responsible for shaping beach deposits. However, due to changes in coastline orientation and inner shelf gradient, beaches along the shoreline will present differences in the degree of exposure to prevailing hydrodynamic factors. Therefore, the actual environmental response will depend on the nature and level of energy inputs, and the physical characteristics of the coastal zone (Short and Hesp 1982). Some beach landforms may persist over time, but they are the exception since most beach geomorphological features are ephemeral due to dynamic action and continuous changes.

Almost all unconsolidated materials that can be transported by waves, currents, wind or glaciers can form beaches. Regarding the size of the constituent sediments, beach deposits in the studied areas have various compositions, from fine sands up to gravel and boulders, whether from marine, terrigenous or biogenic sources. Generally, beach deposits are a combination of the said sources, constantly reworked by the abovementioned transportation agents within regional coastal energy settings.

Fig. 1 Intertidal shipwreck site under a storm surge in Southern Brazil (Photography: Rodrigo Torres)

Coastal processes in the southwest Atlantic are closely related to seasonality of weather patterns. During summer months, when the weather is generally calmer, waves approach the shoreline in more regular swells, carrying in and accumulating sand right above the swash zone, thus resulting in a beach profile which is more concave upwards with a broad berm and steeper beach face (Fig. 2). During winter and mid-seasons, high energy events such as cold fronts, extra-tropical cyclones and storm surges are more frequent, and steep waves mobilize great amounts of sediments in short-term episodic events as they reach the coast. The typical profile response is to reclaim sand deposited on the backshore under previously calmer conditions, carrying it away seaward, where it will be reworked by waves and coastal currents into characteristic longshore bars.

The maximum horizontal and vertical reaches of these disturbances, both seaward and landward, define what is called the *active zone* of the beach profile. Therefore, shipwrecks and beach deposits in this active zone will then be constantly reworked by breaking waves, coastal currents and eolian processes, until long-term changes in environmental conditions come to alter the relative position of the active zone.

In this regard, a useful theoretical concept applied to study sedimentary processes in sand beaches is the idea of *sediment budget*. It is basically an application of the principle of conservation of mass to the littoral sediments and, in practice, consists of estimates of the principal sources of sediment input (credits) and sediment losses (debits) for a stretch of shoreline (Komar 1983: 18). The resulting balance is an indication of the net state of the studied unity, reflected in areas of local beach erosion,

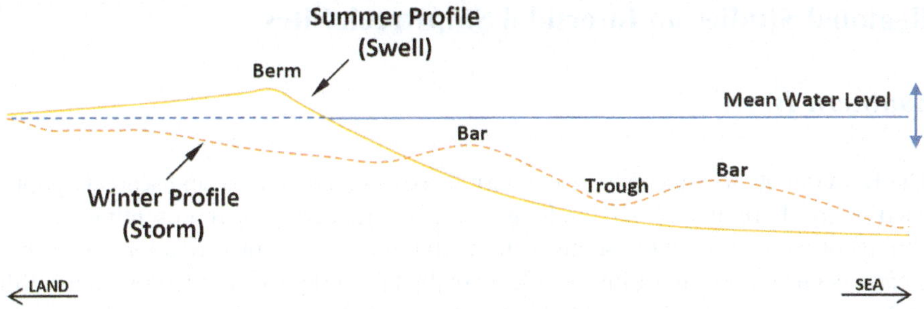

Fig. 2 Typical beach profile responses under winter/storm and summer/swell conditions (modified from Komar 1983: 7)

accretion or stability (steady-state). As a consequence, within historical periods of time, a coastline is said to be *prograding*, when long-term positive sediment balance results in net sediment accumulation and seaward beach growth, whereas it is said to be *retrograding*, when persistent sediment deficits result in a landward retraction of the shoreline.

From a morphodynamic standpoint, beach profiles are controlled by changes in the wave regimes, variations in sediments size and composition, and the interaction between waves and sediments that gives rise to sediment transportation processes. The beach slope increases with grain size and coarse sand and gravel beaches tend to have steeper slopes than fine sandy ones. Differences in slope are related to the greater permeability of the larger grain size sediments (Bagnold 1940; Bascom 1951).

Beaches and dunes of coastal areas are part of a sedimentary system that feeds on the availability of sand and dominant marine and eolian process. Coastal dunes and beaches interact almost continuously through the exchange of sediments, which can occur gradually between one season to another, or more suddenly during highly energetic events, such as under reach of stormy waves and severe winds. Sand dunes are very dynamic coastal features, susceptible to changes in shoreline dynamics, and also vulnerable to anthropic action.

Clearly, a systematic understanding of the regional factors controlling spatial-temporal positioning of the active zone of beach profiles along the shorelines and its local variations is of fundamental importance to the study of beach stranded shipwrecks. Thus, the following section will present our current knowledge regarding the geoarchaeology of intertidal shipwreck sites in southern Brazil, Argentina y Uruguay, in order to highlight regional themes and define the gaps which need to be filled towards considering a more systematic macro-regional approach.

Regional Studies on Intertidal Shipwreck Sites

Southern Brazil

The Rio Grande do Sul coast, in southern Brazil, is an extensive and scarcely populated sandy shore, nearly 620 km long, home of stranded ships representing various periods of navigation in the southwestern Atlantic (Fig. 3). Investigations suggest a greater occurrence of maritime accidents in the last 200 years, especially in the 19th century, due to the losses of merchant sailing and steam vessels engaged in newly opened trading routes between Europe, North America, Southern Brazil, and River Plate provinces.

Fig. 3 Map of South America indicating the study areas mentioned in this chapter

In the course of the "Shore Stranded Shipwrecks of Southern Brazil project" (Torres 2005, 2015), several field trips have been carried out to record and monitor shipwreck sites along the southern and central shores of the study area. Twenty-one sites showing evidence of stranded ships from different periods, construction materials and degrees of preservation have been found along the beach, partially buried in the sand, in the surf, swash zone or high up near the dune field above the high mean water mark.

Together with the information derived from fieldwork, historical documents, newspapers, maritime accidents and shipwreck accounts were compiled with the focus on 19th century. Information on 240 maritime accidents were thus entered in a database, and profiled by: Date of loss, Ship's name, Rigging type, Nationality, Accident location and Cause of loss, plus other useful information. Data from fieldwork and literature review were entered in a GIS Database, allowing for a comprehensive view and management of the archaeological record along the southern Brazilian coast (for details, see Torres 2015).

In 2011, a long term monitoring experiment was set at a site in the central shore (NAV Lagoa do Peixe site), aiming at providing an empirical basis for future modeling of formation processes. Two benchmarks were installed at the site and five control points were established on each structure, allowing for tridimensional surveying of wreck features in relation to the beach's profile changes. In addition, in 2013 two partial excavations were conducted at NAV Lagoa do Peixe and NAV Inedito sites, in order to record shipbuilding features and furnish in situ observations regarding the archaeological record in dynamic settings (Torres 2017).

Furthermore, two cases-studies of maritime accidents in the study area were examined, one in 1861, when the losses of 3 ships were reconstructed based on contemporary newspapers, and a more modern case, which happened in 2003, with the loss a fishing vessel. The case-studies provided useful insights to discuss the specific characteristics of the formation processes of the archaeological record, particularly in terms of the associations between maritime accidents and regional weather patterns, human responses to wrecking events and their signatures in the archaeological record.

Research has shown that beach stranding was the most common cause of maritime accidents in the coast of Rio Grande do Sul during the 19th century. Database records indicate that the majority of accidents correspond to ships stranded on the sand banks at the entrance to the port of Rio Grande (56%). In fact, the difficulties in crossing the Rio Grande bar were legendary among mariners, travelers, and consignees interested in the trade with southern Brazil. The remaining registered accidents (44%) happened along the other areas of the godforsaken sandy stretches of coastline.

From a geomorphological standpoint, the Rio Grande do Sul seashore is a typical sandy barrier-island. The beaches in the barrier experience both cyclical (seasonal) and non-cyclical (meteorological) events of natural flooding, burial and exposure, with significant implications for the preservation of the studied sites, as well as for the distribution of artifacts and the interpretation of archaeological data. Local responses to environmental forcing factors, however, varies along the barrier, due to changes in coastline orientation, inner shelf bathymetry, sediment characteristics, among other aspects, thus presenting differential exposures to prevailing hydrodynamics (Torres 2015: 70).

Although at this point the accumulated knowledge is not sufficient to establish a comprehensive formation processes model, some useful understanding has started to emerge from our research. The long term monitoring experiment, when examined together with field observations, has clearly indicated that the ship structures can move substantially when situated within the active zone of the beach profile, whereas other ship parts in the same site, but situated higher up in the backshore, tend to a lesser degree of horizontal displacement (Torres 2015: 110). What still not yet fully understood, though, is the range of possible responses to changing environmental conditions, and the vertical displacement of shipwreck features and artifacts within the beach profile.

A model proposed by McNinch et al. (2001, 2006) for shallow energetic nearshore settings indicates that wreck features in highly dynamic settings will normally settle into the sand via episodic scour processes, favoring preservation of heavier artifacts and hull parts. The authors observed that depending on site hydrodynamics and underlying geology, heavy features could in fact settle through several meters of substrate when in highly dynamic depositional environments. It suggests potential preservation of artifacts in deeper strata, thus providing a starting point for future probing and understanding of what is called in the specialized literature as depth of disturbance (Williams 1971; Fucella and Dolan 1996; Anfuso et al. 2000).

Nevertheless, the actual survival of an archaeological record will depend on an array of cultural and natural factors involved in pre-depositional, depositional and post-depositional events, which needs to be established in the lights of particular historical, cultural and environmental frameworks. Our research in southern Brazil has indicated that the initial creation of the archaeological record is highly influenced by human attitude towards risk, as well as those circumstances involved in the wrecking process. More properly, the wrecking event should be considered as the time span extending from the moment when preparations are made in anticipation to the accident, until the moment when the ship is abandoned.

The case studies revealed that in the course of the unfolding chain of events, artifacts were jettisoned and ship structures broke apart as a result of attempts to avoid the accident or due to contact with breaking waves. Depending on the materials density, artifacts either sunk or floated away, being eventually deposited elsewhere, often a nearby beach. Moreover, attempts for salvaging ship parts and cargo typically started right away, during the event and/or in the following weeks after stranding, an attitude that often continues throughout the life history of any accessible wreck site. These observations are in line with Gibbs (2006) approach on cultural site formation processes, and Muckelroy's (1978: 157–175) "Extracting Filters and Scrambling Devices" in effect during the process of wrecking.

In an interesting study, Delgado and Murphy (1984) proposed a classification scheme for the wrecking processes of beached stranded shipwrecks, which includes three categories: buoyant hull sites, buoyant hull fracture, and buoyant structure sites. Buoyant hull sites are those in which a ship comes ashore and settles into the sand, relatively intact. Buoyant hull fracture sites occur when the hull of a ship comes

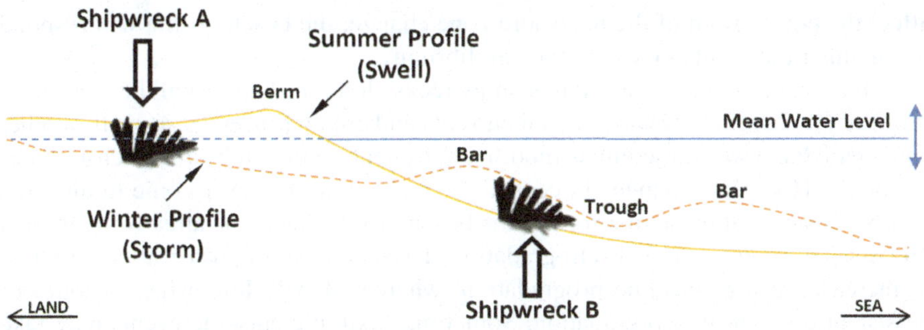

Fig. 4 Idealized shipwreck burial/exposure responses to changes in beach profile (modified from Komar 1983: 7)

ashore intact, but it breaks apart on the beach and is dispersed by the surf. Buoyant structure sites, in turn, are formed when a ship breaks apart offshore and washes onto the beach in pieces.

The case study of the loss of a fishing vessel which happened at the southern shore in 2003 revealed a good fit for the buoyant hull fracture type. The keel and lower hull, being stuck in the sand banks underwater, caused the hull upper works to snap in the beating waves. These larger pieces were further broken down, following the initial snap, until they reached an initial resting position of equilibrium in relation to prevailing hydrodynamic conditions. Since these weather systems are often coupled with storm surges, sometimes lasting up to several days, the initial hull resting position may be potentially higher up in the beach profile due to water piling up against shore, than it would otherwise, if stranded in calmer conditions. Other factors, such as the ship's draft, for instance, may also affect this initial resting place.

Following previously discussed theoretical responses for beach profiles during changing hydrodynamics, two idealized wrecking scenarios have been considered for the Rio Grande do Sul coast (Fig. 4).

During highly energetic weather and sea settings, when waves packed with energy reach the shore and waves sets up over the beach face mobilizing great amounts of sediments, the typical profile response is to reclaim sand deposited in the backshore under previously calmer conditions, carrying it away seaward forming characteristic longshore bars (dashed profile line). Under such conditions, shipwrecks in position A would likely be exposed, while those found in position B are more likely to be partially or totally covered by sediments.

During calmer conditions, waves approaching in more regular swells drive the underwater sandbars ashore, carrying in and accumulating sand above the swash zone. This results in a beach profile which is more concave upwards with a broad berm and steeper beach face (solid profile line). Under these conditions, sediments are carried in from underwater towards dry land, and shipwrecks placed in the backshore (position A) can be expected to be partially or totally covered by sediments, while those found under water (position B) are exposed. In practice, as the forces that affect the equilibrium of the nearshore zone change, the beach profile will respond in a manner that tends to restore the equilibrium.

In the long run, it is expected that shipwrecks along the barrier will be constantly reworked by breaking waves, coastal currents and eolian processes, as well as other ephemeral but powerful sediment mobilization agents in the study area, such as dunes washouts. Historical changes in coastal dynamics may however come to alter the relative position of the active zone of the beach profile due to the already mentioned processes of progradation or retrogradation. Thus, chances of preservation will tend to increase during coastline progradation, whereas it will diminishes in zones of persistent erosion or retrogradation. Sometimes extreme episodic events may take the hull to a position beyond the reach of regular seasonal disturbances, either towards onshore or offshore. In other situations no integral fragment is left.

The scenarios discussed above are typical and correspond to highly simplified geomorphological responses in exposed sandy shores where morphodynamics are primarily dominated by waves, such as in southern Brazil. Given the complexity of the depositional environment, the totality of factors controlling vertical and horizontal distribution of cultural material from beach stranded shipwrecks are still unclear, so a complete model of formation processes cannot yet be proposed. The development of more robust theoretical contributions in this field must come from an approach capable of accounting for long-term trends in the displacement of ship remains in relation to environmental parameters.

Argentina

Ever since the discovery of the Strait of Magellan by Europeans, in 1520, the exploration, exploitation and colonization of the current territory of Argentina was ruled by navigation, particularly around the southernmost parts of the country following the main routes that connected the Atlantic and Pacific oceans. By the 19th century, and until the opening of the Panama Canal in 1914, the magnitude of vessels from the northern hemisphere, Europe and North America was complemented by other watercraft, usually of small size and aimed at regional navigation, constructed in the Parana-Plata basin with local timbers (see Castelli 2017 for an in-depth discussion). Whatever their geographic origin, vessels of different type and size navigated along the coast and accidents were frequent. As it is the case with all shipwreck sites, the archaeological remains of such mishaps represent a valuable testimony of multiple cultural aspects of the past, from life on board to the functions of the vessels and their construction techniques, among others.

From the beginning of the discipline of maritime and underwater archaeology in the country the main research efforts were devoted to a fully submerged shipwreck site, the British sloop of war HMS Swift lost in Puerto Deseado, Patagonia in 1770 (see Elkin et al. 2011, for a comprehensive account of the project). Over the years, however, land coastal surveys have also been conducted with different objectives and at different scales, to which this section will be referred. Some of the projects, for instance, had a component of surveying large extensions of coastline for in situ recording of shipwreck remains, aiming at assessing the nature of the maritime archaeological record prior to selecting case studies for further site-specific research. The latter could include determining the geographical and chronological context of each site, understanding their natural and cultural formation processes or their function, conducting technological analysis, or establishing management priorities.

Good examples of large-scale surveys and interpretations of coastal shipwreck remains are the studies conducted in three of the areas indicated in Fig. 3: Puerto Madryn and Peninsula Valdés, in Chubut Province (Elkin et al. 2015), the coast of Monte León National Park, in Santa Cruz Province (Grosso et al. 2015), and Peninsula Mitre, on the Atlantic coast of Tierra del Fuego Province (Elkin et al. 2017).

Materials were mostly found situated in intertidal and supratidal environments, sometimes with underwater components as well. In many cases the archaeological record was disarticulated and, particularly in the southernmost areas of Patagonia and Tierra del Fuego, scattered along extensive sections of the coastline.

A promising approach aimed at evaluating correlations between the fragmentation or disarticulation of shipwrecks and the specific depositional environment in which they are placed is presented in Grosso et al. (2016), considering the three areas mentioned above. Representative profiles of the types of coastal environments in which findings are located were characterized, including data from substrate type and slope. Different categories referring to the degree of hull fragmentation and disarticulation were created to see if it was possible to identify association patterns in between fragmentation levels and the type of depositional environment where the remains were placed.

The coherence of each category was analyzed using the following indicators: material of structural elements, i.e. wood or metal (either iron or steel), dimensions of individual elements and structures, and construction system. Subsequently, more specific analyzes were carried out (taxonomic identification of the wood, metallographic analyses) and the data were compared with classification rules, insurance records and other sources. In this way the size, function, construction material, propulsion system, chronology and cultural origin of the vessels could be interpreted. A minimum number of shipwrecks was also defined. The results pointed to an assembly made of vessels with different construction typologies (wooden hull, metal or mixed), of medium to large size (between 200 and 2,000 tons) and propelled mostly by sail. The time period ranges from the beginning of the 19th century to the beginning of the 20th century. On the other hand, there were also numerous elements that could not be linked to any of these vessels.

Subsequent studies conducted in the Atlantic coast of Tierra del Fuego (Elkin et al. 2017) resulted in a similar classification of shipwreck remains, as the ones presented in Grosso et al. (2016), although in this case the archaeological material is not only considered from the fragmentation point of view but also in terms of the degree to which the hull is represented.

The main point worth remarking is that these large-scale approaches in shipwreck archaeology in Argentina are beginning to reveal some recurrent patterns, providing a basis to predict the probabilities of finding different types of archaeological shipwreck materials in a given environment.

As mentioned earlier, some site-specific studies were also conducted in different parts of the Patagonian coast, relating to intertidal or semi submerged shipwreck remains. The sites subjected to the most thorough surveys and researches were the "Presidente Roca", a large metal hulled steamer which caught on fire in 1909 in Peninsula Valdés (Elkin et al. 2015; Gutiérrez and Elkin 2010; Gutiérrez 2013, 2016), the site called Bahía Galenses in Puerto Madryn, which represents a wooden-hulled 19th century merchantman (Elkin et al. 2015; Murray et al. 2009), and the remains of the "Hoorn", a Dutch *jaacht* of the Le Maire and Schouten expedition, lost in the Deseado estuary, Santa Cruz Province, in 1615 (Murray et al. 2008, 2009). In all of the abovementioned sites, different variables were taken in account regarding natural formation processes.

Quite recently, another maritime archaeology project began to be conducted further north of the areas previously mentioned. The studied area is situated at Cabo San Antonio, located about 300 km south of the city of Buenos Aires, where in the 19th century it had a significant economic boom that generated an increase in navigation traffic. A wide range of shipwreck remains were found in different depositional contexts, such as inland over the dunes, in the intertidal zone and submerged, as well as in different degrees of integrity: some of them still with complete structures, while others were fragmented (Pisani 2006). Given the lack of homogeneity in the maritime archaeological record in Cabo San Antonio, and the need to propose regional investigations, the initial approaches involved understanding sediment dynamics through geo-archeological surveys, considering a particular shipwreck located at the intertidal zone: the Costa del Este shipwreck (Fig. 5), a small wooden, apparently single-masted vessel attributed to the 19th century (Elkin et al. 2016a).

Beach profiles were considered a useful tool to study sedimentary processes, since the coast of Buenos Aires province is characterized by dynamic sandy beaches, especially in the Cabo San Antonio, where they are affected by both cyclic (seasonal) and non-cyclic (meteorological) changes. A beach profile registers the beach face as shaped by wave dynamics and sediment grain size (Komar 1983). According to the project goals, beach profiles were taken in different seasons, accompanied by collection and analysis of sediment samples. An annual monitoring of the waves (period and height) and winds (direction) was also carried out. The results showed that there were seasonal variations in the profiles: during the summer period the profile corresponds to one formed by waves of lower energy, with the presence of a channel and a bar, where the grain size was variable. During the winter period, instead, a flat profile, formed as a result of waves of higher energy, can be seen (De Hoz 2017; Elkin et al. 2016b).

These studies about sedimentary dynamics are also of great importance for heritage management, as the prediction of burial/exposure of vessel remains is important when considering public access to the foreshore maritime heritage. Even if shipwreck sites on land are partially covered, they are more accessible to the public than any fully submerged site simply because the visitors are not restricted to the diving community.

As already said, the study of the Costa del Este site was a first approach to understand sediment dynamics in the Cabo San Antonio area. As De Hoz (2017) observed, from the point of view of considering more systematic regional investigations, it would be interesting to group the studied shipwreck sites into categories comparable to those used in the Patagonian and Tierra del Fuego sectors, where sites were classified according to their degree of fragmentation and hull representation. Moreover, a broader systematic study should consider establishing the relationships between shipwrecks, type of substrate and the section of the beach in which they are located, whether supratidal, intertidal or submerged, in correlation with more specific environmental variables, such as littoral currents and sediment grain size.

As the interest to study archeological record located in the intertidal zone grows, so does the need to systematize survey methodologies and continue developing large scale studies (Grosso et al. 2018, in press). It highlights the need to implement different ways to approach the archeological record, incorporating substantive knowledge from other disciplines, such as geology, biology, history, geography and archaeology to create a tool for better interpretation and management of the archeological record. In addition, the advance in the knowledge of our intertidal spaces and their interaction with archaeological materials will make it possible to understand the processes linked to the wreck sites formation processes in those environments.

Fig. 5 Costa del Este Shipwreck (Photography: Juan Croce)

Interdisciplinary studies are required for such purpose, and they are already shedding interesting light in that sense (Elkin et al. 2015; Grosso et al. 2013, 2018). In the future, it will be possible to postulate models that contemplate or can quantify some of the already mentioned variables in order to typify formation processes in relation to specific intertidal environments. In this way, the risks to which the materials are exposed can also predicted, thus contribute to better decision making relating to their management and conservation, either by preserving them in situ or by recovering them fully or partially. In fact, two of the nautical coastal archaeology projects mentioned above, the ones that took place in Monte León National Park and the Atlantic Coast of Tierra del Fuego, originated as baseline studies which could be subsequently useful for management purposes. As Oxley quite rightly points out (2016: 215), "good data is essential for good decision making".

Uruguay

The Uruguayan coastline has an approximate extent of 700 km, encompassing various oceanic, estuarine and riverine sub-environments, with different geomorphological and hydrodynamic characteristics. The study area for this paper comprehends two main coastal sectors: the River Plate estuarine sector, which extends 473 km W-E from Nueva Palmira to Punta del Este, and the Atlantic coast sector, extending 233 km mainly SW-NE from Punta del Este to Barra del Chuy, on the border with southern Brazil (Fig. 3).

These waters have been recurrently navigated since the period of European conquest and earlier by pre-Hispanic indigenous peoples, strongly sharing its maritime history with Brazil, to the north, and with Argentina, with which it shares the River Plate navigational lanes. This section thus presents our current knowledge regarding investigations on intertidal shipwreck sites along the Uruguayan Atlantic coast and the outer parts of the River Plate estuary.

Based on an ongoing literature review and archival research, a total number of 1,071 maritime accidents have been registered for the coasts of Uruguay, between the 16 and 20th centuries, including both deep water losses and coastal strandings. Moreover, based on fieldwork and internet research, 34 intertidal wreck sites have been identified in the study area thus far, and a database is being continuously fed with new information by the staff at the Coastal Heritage Research Center of the Universidad de la República, Uruguay.

In general lines, the coast of both sectors is composed of a succession of sandy beaches separated by rocky promontories, which adopt arched forms of variable curvatures as a result of their orientation and exposure to prevailing waves (UNESCO 1979; Gómez 2010). These sandy arches are formed by unconsolidated sediment which is in constant movement between nearshore and foreshore, as well as transported along shore within littoral drift cells forced by waves (Panario and Gutiérrez 2005).

Within a microtidal regime, the main natural acting forces shaping beach environments in the study area are the waves, winds and coastal currents. Wind-driven coastal surges (meteorological tides) play an important role in regional coastal processes in both considered sectors. Wind dominance at these latitudes has a typical sub-tropical bimodal NE-SW oppositional regime, where winds from the southern quadrants are the strongest and prevail during winter and mid-seasons, whereas winds from the northern quadrants are more frequent throughout the year.

The strong, and often severe, winds from the southern quadrants are typically associated with the passage of transient high-pressure systems, cold front intensification and cyclogenetic activity (Gan and Rao 1991). The navigational hazard posed by these winds, regionally called "pampero winds" (SW) and "sudestadas" (SE), are long known by captains and mariners sailing along the coast or in demand of River Plate ports (Lobo and Riudavets 1868: 181–188).

It is understood that coastal processes are of great influence for coastal navigation, now and then, as well as for beach environment variability, thus affecting the occurrence, distribution and survival of the archaeological record on both sectors of the Uruguayan coastline herein considered. Their influence on each sector, however, is dissimilar, mainly due to differences in shoreline orientation and nearshore bathymetry.

While in the Atlantic coast, where the shoreline has a predominant NE-SW orientation, the effects of coastal wind/sea interaction respond mainly to Ekman transport, whereas in the River Plate sector, winds from the southern quadrants strike violently near-normal toward the coast, resulting in a high setup, particularly when superimposed to high astronomical tides, larger and choppy surf zone, and strong coastal currents are formed, affecting sediment balance. However, considering the intricacies in shoreline orientation and exposure in arched beaches, various local responses in beach morphodynamics and sediment balance are to be expected, and more in depth, long term monitoring studies of shipwrecks at individual beaches must be considered.

Some of the identified intertidal wreck sites have already been studied in different degrees. In 2000, archaeologists and staff from the National Heritage Commission studied the remains of a large 19–20th century merchantman at the intertidal zone of Santa Rosa beach, department of Canelones (Buffa and Cordero 2010). The study revealed three large preserved hull structures, spread along the beach arc. The investigation focused on documenting shipbuilding features of the excavated hull structures, aiming at chronological and typological identification, understanding the environmental processes responsible for site formation and the artifact distribution in arched beaches.

Between 2004 and 2009, in the framework of the academic activities of the Underwater Archaeology Program of the Universidad de la República, a series of investigations concentrated on recording and interpreting the underwater cultural heritage began in Piriápolis, Department of Maldonado (Pereira 2005; Lezama et al. 2006, 2009). The interventions at the Vértigo site revealed highly disintegrated archaeological remains of one or more shipwrecks, deposited at a shallow setting on a rocky promontory and its adjacent beach (Keldjian 2014).

Although no integral hull remains were observed, artifacts and hull features found on the underwater as well as in the sub-aerial components of the site are storytelling. Underwater, the survived record was composed essentially by iron and lead ordinance, including a cannon, cannon balls and shots, as well as iron fasteners and knees, preserved within concretions wedged in between the rocks. The laboratory conservation of the cannon, carried out between 2005 and 2006, also revealed unspoiled organic cordage still attached to its cascabel, preserved under thick concretion layers. Systematic beach surveys, in the other hand, identified 104 artifacts of which less than 25% could be somewhat associated with shipwreck remains, mainly composed of iron concretions, riddled timbers with iron oxide stains and coal fragments.

Several other artifacts and fragments allegedly from the wreck site, previously collected underwater and along the beach by local townspeople, were also recorded. This assembly was more varied and included, apart from shots of different sizes, also fragments of lead and copper sheathing, copper and bronze spikes, and two late-18th century copper coins. Confronting the archaeological, historical, and oral records, the researchers narrowed down the wreck identification to three possible candidates: the Spanish frigate "Nuestra Señora de la Estella", the English frigate "Duque of Clarens", and the French brigue "Le Consolateur", lost in 1794, 1805 and 1808, respectively.

As observed by Stewart (1999: 582), shallow water waves and currents, eroding and transporting artifacts and features away from the original wreck site, pose a hard challenge for the interpretation of shipwrecks and artifact patterning along highly energetic shorelines. The work at the "Vértigo" site, however, have shown that sound, multicomponent foreshore systematic surveys, particularly within the context of an enduring academic research, can produce encouraging results, both in terms of archaeological knowledge, as well as in terms of understanding how local communities interact with their underwater cultural heritage.

Many other scientific shallow water beached wreck surveys have been carried out in Uruguay since the 2000s, but few have yet been formally published, which makes it difficult for assessing their contribution to a more systematic approach on the problem under examination. Furthermore, from mid-1980s to 2007, several commercial salvage permits were issued, resulting in the excavation of dozens of high profile shipwreck sites located in various parts of the Uruguayan coast of the River Plate and adjacent Atlantic Ocean. Unfortunately, very little documentation was produced during these interventions, and no integral artifact collections survived.

Recently, new efforts for systematic studying of intertidal archaeological sites are being devised by students and researchers associated with the Coastal Heritage Research Center at the Universidad de la República. Incorporating broader conceptual and methodological approaches, a project called "The Archaeological Record Integrated: assessing the underwater continuity of coastal sites" seeks at enlarging the perception of coastal sites as an integrated land/water archaeological space, based on the study of two sites with different chronological, typological and environmental settings, being one a riverine site at the Río Negro department, and another at an outer section of the River Plate estuary, at Maldonado Bay. Moreover, in the framework of the project "Digital 3D Technologies applied to the Study, Management, and Public

Awareness of the Coastal Heritage in Uruguay", some of the authors have proposed a systematic study of stranded shipwreck sites, using 3D recording and visualization technologies for assessing formation processes and long term monitoring, as well as public sharing of the coastal and maritime heritage in Uruguay.

Final Considerations

Beached shipwreck studies in South America are quite varied and have been driven by different circumstances and goals (Elkin 2011). As Murphy (1983) already mentioned, beached shipwreck sites are important, therefore they cannot be dismissed on the assumption that the information they can contain is poor. That would be a serious mistake that will compromise the amount of data ultimately obtainable from the shipwreck, and which can be related to construction techniques, site formation processes and many other aspects of past human life and the archaeological site itself.

However, this data will only emerge from rigorous investigations, specific to the archaeological record under consideration. For such an approach, it is necessary to consider the physical setting and dynamic processes which are taking place, but also anthropogenic causes, as not only environmental factors play an important role but also human behavior involving activities such as salvaging, pillaging or simply souvenir collecting participate in the shaping of the archaeological record. It is evident that interdisciplinary approaches considering beach geomorphology, sedimentology, coastal oceanography and meteorology, botanical and metallurgical analysis, human behavior, among other aspects, will yield the richest information.

In order to establish adequate criteria for selecting specific sites, or specific areas to conduct research, good shipwreck databases are needed. In Uruguay, CIPAC is currently carrying out Geographic Information System (GIS) files with the all the information on shipwrecks collected so far. First, the information is obtained from different sources: cartography, press, custom-house, logbooks, photographs, etc. A second goal is to interpret the documentation in relation to the geographical location, and finally, to generate graphic representation in the GIS file with the information for each case. The aim of this systematization is to provide the information available to users, through the web—GIS interface.

The Underwater Archaeology Program (PROAS) in Argentina has also been developing a shipwreck database using sources from the Navy, the Coast-Guard and other official repositories, sometimes complemented with more informal–not necessarily less reliable- sources such as contemporary personal communications, photographs, etc., but the incorporation into a GIS is in its initial stages.

In Brazil the Navy has a shipwreck database related to an Underwater Archaeology Chart but it has restrictions for public access (Flavio Calippo, pers. comm.). The National Historic and Artistic Heritage Institute (IPHAN) has a GIS database of archaeological heritage organized by region which is accessible to the public in their website (iphan.gov.br). The database is almost exclusively conformed by land sites.

For comparative purposes it would be extremely useful if the shipwreck databases of different countries shared at least the same basic variables. Logically, historical accounts of accidents would form its basis, with a gradual incorporation of archaeological data of sites with confirmed existence and location, including of course intertidal sites.

Although shipwreck archaeology has come a long way in recognizing the potential of less preserved shipwreck sites, there is still some latitude for methodological developments. High energetic and dynamic shipwreck sites demand specific field procedures to account for the study of long-term patterning of the material in relation to the environment. One conclusion of the work in Brazil, surely applicable elsewhere, is that the more dynamic the environment, the greater the need for long term monitoring experiments. This, however, is more difficult to implement in remote places (such as Patagonia, Tierra del Fuego, or the southern Brazilian shore) so alternative resources should be sought.

Another useful methodological tool, as shown in the Brazil section of this chapter, would be to incorporate case studies based on modern maritime accidents where there is considerable information to better understand the early stages of wrecking process. In short: monitoring, formation processes experiments and modern shipwrecks case-studies—ideally in a systematized and broader macro-regional large, could help establishing parameters and variables for modeling formation processes and decision making.

One conclusion of this review is that Uruguay has a lot of potential for the study of beached shipwrecks (and shipwrecks in general) particularly now that contracts for commercial exploitation of such sites has been stopped. After decades of being constrained by such circumstances, despite of which there has been considerable progress in research, management and capacity building, maritime archaeologists working in the country can now start selecting what and how to study. Hopefully, beached shipwrecks will be increasingly included in future projects.

Clearly, the approach given to the study of coastal shipwreck sites in the area considered in this chapter is far from homogeneous, and it relates to variation in the development of the discipline of maritime archaeology in the different countries involved and the goals and circumstances of each case study. However, as seen, there seems to be a sufficient sample of archaeological record covering this large oceanic and estuarine coastline in order to begin comparative studies.

For example, it might be worth evaluating if the correlation observed in shipwreck remains in Patagonia and Tierra del Fuego between ship integrity and environmental location can also be applied to those located in Southern Brazil and Uruguay. Or, if the GIS database which is being used in Uruguay can be successfully integrated to the databases used in Argentina, perhaps aiming at regional management approach of the River Plate.

The degree of preservation of shipwreck remains along the whole area should not be an obstacle for rich comparative studies along those lines. Furthermore, wherever the remains of a ship's hull are observed, the internal logic of nautical architecture warrants a high level of informational coherence, so that, in the words of Steffy (1996: 6): "How well must a ship or a boat be preserved before a thorough study becomes necessary? There is no lower limit".

References

Adams J (2009) Comments on Keith Muckelroy: methods, ideas and maritime archaeology. J Marit Res 4:83–85

Anfuso G, Gracia FJ, Andrés J, Sánchez F, Del Río L, López-Aguayo F (2000) Depth of disturbance in mesotidal beaches during a single tidal cycle. J Coast Res 12(2):446–457

Ascher R (1961) Analogy in archaeological interpretation. Southwest J Anthropol 17:317–325

Bagnold R (1940) Beach formation by waves: some model experiments in a wave tank. J Inst Civil Eng 15(1):27–52

Bascom W (1951) The relationship between sand-size and beach-face slope. Trans Am Geophys Union 32:866–874

Bass G (1966) Archaeology under water. Thames and Hudson, London

Bass G (1972) A history of seafaring based on underwater archaeology. Walker and Company, New York

Bass G (1983) A plea for historical particularism in nautical archaeology. In: Gould R (ed) Shipwreck anthropology. University of Mexico Press, Albuquerque

Binford L (1964) A consideration of archaeological research design. Am Antiq 29:425–441

Binford L (1981) Behavioral archaeology and the 'Pompeii premise'. J Anthropol Res 37:195–208

Binford L (1983) Pursuit of the past: decoding the archaeological record. Thames and Hudson, London

Buffa V, Cordero A (2010) Arqueología marítima en el fortín de Santa Rosa: el potencial de los hallazgos costeros. In: Oliva F et al (eds) Arqueología Argentina en los inicios de un Nuevo Siglo. Tomo 3, 1st edn, Laborde Libros, Rosario, pp 293–306

Burke H, Smith C (2004) The archaeologist's field handbook. Allen & Unwin, Crow's Nest

Castelli A (2017) Obra viva: estudio de la construcción y contexto de uso de una embarcación naufragada en el río de la plata en el siglo XIX. Unpublished graduation thesis in Anthropology, with major in Archaeology. Facultad de Filosofía y Letras, Universidad de Buenos Aires. MS on file, Universidad de Buenos Aires

Clarke D (1973) Archaeology: the loss of innocence. Antiquity 47:6–18

Delgado J (1984) Erosion-exposed shipwreck remains, winter 1982. In: Murphy L (ed) Submerged cultural resources survey: portions of point Reyes national seashore and point Reyes-Farallon islands national marine sanctuary. National Park Service, Santa Fe, pp 179–185

Delgado J (1985) Skeleton in the sand: documentation of the environmentally exposed 1856 ship King Philip, San Francisco. In: Johnston P (ed) Proceedings of the sixth conference on underwater archaeology, Minnesota

Delgado J (1997) Beached shipwreck sites. In: Delgado J (ed) Encyclopaedia of underwater and maritime archaeology. British Museum Press, London, pp 57–59

De Hoz D (2017) Procesos de formación naturales en sitios de naufragio en la costa bonaerense. El caso de los restos de una embarcación en Costa del Este. Unpublished graduation thesis in Anthropology, with major in Archaeology. Facultad de Filosofía y Letras, Universidad de Buenos Aires

Dumas F (1962) Deep water archaeology. Routledge, London

Dumas F (1972) Ancient wrecks. In: Underwater archaeology, a nascent discipline. UNESCO, Paris

Duncan B (2000) Signposts in the sea: an investigation of the shipwreck patterning and maritime cultural landscapes/seascapes of the Gippsland Region, Victoria. Unpublished British Society of Science (Hons) thesis, James Cook University, Townsville, Australia

Dunnell R (1992) The notion site. In: Rossignol J, Wandsnider L (eds) Space, time and archaeological landscapes. Plenum Press, New York, pp 21–39

Elkin D (2011) In: Ford B, Hamilton DL, Catsambis A (eds) Shipwreck archaeology in South America. Oxford University Press, New York, pp 685–707

Elkin D, Murray C, Bastida R, Grosso M, Argüeso A, Vainstub D, Underwood C, Ciarlo N (2011) El naufragio de la HMS Swift (1770). Arqueología marítima en la Patagonia. Vázquez Mazzini Editores, Buenos Aires

Elkin D, Murray C, Grosso M, Gutiérrez G, Trassens M, Bastida R (2015) Investigaciones interdisciplinarias en sitios arqueológicos de naufragio en el Golfo Nuevo (Provincia Del Chubut): Primeros resultados. Cuadernos del Instituto Nacional de Antropología y Pensamiento Latinoamericano 24(1):21–40

Elkin D, Mundo I, Croce J (2016a) Primeras aproximaciones al Pecio de Costa del Este (Partido de la Costa, Provincia de Buenos Aires, Argentina). Cuadernos del Instituto Nacional de Antropología y Pensamiento Latinoamericano, Series Especiales 2(4):191–206

Elkin D, Kokot R, De Hoz D (2016b) Procesos de formación de sitios de naufragio en la interface playa-duna. Primeras aproximaciones referidas al litoral atlántico bonaerense. Libro de Resúmenes del XIX Congreso Nacional de Arqueología Argentina. San Miguel de Tucumán

Elkin D, Murray C, Grosso M (2017) Arqueología de naufragios históricos en la costa atlántica fueguina. In: Vázquez M, Elkin D, Oría J (eds) Patrimonio a orillas del mar - Arqueología del litoral atlántico de Tierra del Fuego, Editora Cultural de Tierra del Fuego, Ushuaia, pp 207–237

Frost H (1962) Submarine archaeology and Mediterranean wreck formations. Mar Mirror 48:82–89

Fucella J, Dolan R (1996) Magnitude of subaerial beach disturbance during northeast storms. J Coast Res 12(2):420–429

Gan M, Rao V (1991) Surface cyclogenesis over South America. Am Meteorol Soc 19:1293–1302

Gibbs M (2006) Cultural site formation processes in maritime archaeology: disaster response, salvage and Muckelroy 30 Years on. Int J Naut Archaeol 35(1):4–19

Gómez M (2010) Los asuntos claves para el manejo costero integrado en Iberoamérica: Uruguay. In: Barragán Muñoz J (ed) Manejo Costero Integrado y Política Pública en Iberoamérica: Un diagnóstico. Red IBERMAR, Cádiz, pp 261–290

Grosso M, Trassens M, Murray C, Bastida R (2013) Relevancia de la matriz sedimentaria en los sitios arqueológicos subacuáticos. Revista de Arqueología Americana 31:205–246

Grosso M, Elkin D, Murray C, Castro M (2015) Evidencia arqueológica de naufragios en zonas costeras: el caso del Parque Nacional Monte León (Provincia de Santa Cruz). Magallania 43(2):93–118

Grosso M, Elkin D, Murray C, Bastida R (2016) El registro arqueológico de naufragios históricos: consideraciones metodológicas sobre prospecciones costeras en el Atlántico Sur argentino. In: Mena F (ed) Arqueología de Patagonia: de Mar a Mar. Ediciones CIEP-Ñire Negro, Santiago de Chile, pp 90–99

Grosso M, Trassens M, Murray C, Bastida R (2018) Caracterización de los ambientes intermareales y su aplicación en el estudio del registro arqueológico en el litoral marítimo argentino. Cuadernos del Instituto Nacional de Antropología y Pensamiento Latinoamericano 27(2). Buenos Aires (In press)

Gutiérrez G (2013) Arqueología de pecios modernos: Estudio tecnológicos y procesos de formación en el sitio vapor Presidente Roca (Península Valdés, Provincia del Chubut). Unpublished graduation thesis on Anthropological Sciences, Facultad de Filosofía y Letras, Universidad de Buenos Aires

Gutiérrez G (2016) Procesos de formación de sitio en un naufragio de Siglo XX. El caso del vapor Presidente Roca (Península Valdés, Chubut). La zaranda de Ideas. Revista de Jóvenes Investigadores en Arqueología 14(2):131–150

Gutiérrez G, Elkin D (2010) Barcos a vapor en la Patagonia. Primeras aproximaciones al naufragio del Presidente Roca (Península Valdés, Chubut). In: Bárcena R, Chiavazza H (ed) Arqueología Argentina en el Bicentenario de la Revolución de Mayo. VII Congreso Nacional de Arqueología Argentina. Universidad Nacional de Cuyo, Mendoza, pp 2039–2044

Harpster M (2009) Keith Muckelroy: methods, ideas and maritime archaeology. J Marit Res 4:67–82

Harris E (1989) Principles of archaeological stratigraphy, 2nd edn. Academic Press Limited, San Diego

Keith M (2016) Site formation processes of submerged shipwrecks. University Press of Florida, Gainesville

Keldjian E (2014) Arqueología marítima y subacuática: El registro arqueológico y los documentos relativos al sítio Vértigo. Taller II de Arqueología. Facultad de Humanidades y Ciencias de la Educación, Universidad de la República, Montevideo

King C (1972) Beaches and coasts. Edward Arnold, London

Komar P (1983) Beach processes and erosion, an Introduction. In: Komar P (ed) CRC handbook of coastal processes and erosion. CRC Press Inc., Boca Raton

Lezama A, Pereira V, Fernández E (2006) Informe final, Investigación arqueológica del pecio de la nave capitana de Martim Affonso de Souza (1531). Programa de Desarrollo Tecnológico, Ministerio de Cultura, Montevideo

Lezama A, D'Ambrosio L, Saccone E, Keldjian E (2009) Informe final, El Patrimonio Cultural Sumergido en la zona de Piriápolis, Diagnostico y plan de gestión 2007–2009. CSIC, Montevideo

Lobo M, Riudavets P (1868) Manual de la Navegación del Río de la Plata, 2nd edn. Establecimiento Tipográfico Fortanet, Madrid

McCarthy M (1998) Australian maritime archaeology: changes, their antecedents and the path ahead. Aust Archaeol 47:33–38

McNinch JE, Wells JT, Drake TG (2001) The fate of artifacts in an energetic, shallow-water environment: scour and burial of Queen Anne's revenge. Southeast Geol 40(1):19–27

McNinch JE, Wells JT, Trembanis AC (2006) Predicting the fate of artefacts in energetic, shallow marine environments: an approach to site management. Int J Naut Archaeol 35(2):290–309

Muckelroy K (1975) A systematic approach to the investigation of scattered wreck-sites. Int J Naut Archaeol 4:173–190

Muckelroy K (1977) Historic wreck sites in Britain and their environments. Int J Naut Archaeol Underw Explor 6(1):47–57

Muckelroy K (1978) Maritime archaeology. Cambridge University Press, Cambridge

Murray C, Vainstub D, Manders M, Bastida R (2008) Tras la estela del Hoorn; arqueología de un naufragio holandés en la Patagonia. Vázquez Mazzini Editores, Buenos Aires

Murray C, Grosso M, Elkin D, Coronato F, De Rosa H, Castro M, Bastida R, Ciarlo N (2009) Un sitio costero vulnerable: el naufragio de "Bahía Galenses" (Puerto Madryn, Chubut, Argentina). In: Salemme M et al (eds) Arqueología de la Patagonia. Una mirada desde el último confín. Editorial Utopías, Ushuaia, vol 2, pp 1093–1108

Murphy L (1983) Shipwrecks as data base for human behavioral studies. In: Gould R (ed) Shipwreck anthropology. University of New Mexico Press, Albuquerque

Murphy L (1997) Site formation processes. In: Delgado J (ed) British museum encyclopaedia of underwater and maritime archaeology. British Museum Press, London

O'Shea J (2002) The archaeology of scattered wreck-sites: formation processes and shallow water archaeology in western Lake Huron. Int J Naut Archaeol 31(2):21 1–227

O'Shea J (2004) The identification of shipwreck sites: a Bayesian approach. J Archaeol Sci 31:1533–1552

Oxley I (2016) English heritage and shipwreck site formation processes. In: Keith M (ed) Site formation processes of submerged shipwrecks. University Press of Florida, Gainesville, pp 211–234

Panario D, Gutiérrez O (2005) La vegetación en la evolución de las playas arenosas. El caso de la costa uruguaya. Revista Ecosistema 14(2):105–161

Pereira V (2005) Arqueología subacuática: implementación de un sitio escuela Vértigo, Piriápolis, Maldonado. In: 11th Congreso de Arqueología Uruguaya, Salto

Pisani A (2006) La fantasía del naufragio. Author´s Edition, Buenos Aires
Quinn R, Forsythe W, Breen W, Boland D, Lane PJ, Lali Omar A (2007) Process-based models for port evolution and wreck site formation at Mombassa, Kenya. J Archaeol Sci 34:1449–1460
Renfrew C, Bahn P (2008) Archaeology: theories, methods, and practice, 5th edn. Thames & Hudson, New York
Richards N (2002) Deep structures: an examination of deliberate water craft abandonment in Australia. Unpublished PhD dissertation, Department of Archaeology, Flinders University, Australia
Richards N (2003) The role of isolation in cultural site formation: a case study from Strahan, Tasmania. Bull Australas Inst Marit Archaeol 27:77–84
Richards N (2004) The role of geo-politics in cultural site formation: a case study from the Northern Territory. Bull Australas Inst Marit Archaeol 28:97–106
Russel M (2002) Beached shipwrecks from channel Islands national park. Calif J Field Archaeol 29(3):369–384
Schiffer M (1972) Archaeological context and systemic context. Am Antiq 37:156–165
Schiffer M (1976) Behavioral archeology. Academic Press, New York
Schiffer M (1996) Formation processes of the archaeological record. University of Utah Press, Salt Lake City
Schiffer M, Rathje W (1973) Efficient exploitation of the archeological record: penetrating problems. In: Redman C (ed) Research and theory in current archeology. Wiley, New York, pp 169–179
Short A, Hesp P (1982) Wave, beach and dune interactions in Southeastern Australia. Mar Geol 48:259–284
Souza D (1998) The persistence of sail in the age of steam: underwater archaeological evidence from the Dry Tortugas. Plenum, New York
Steffy R (1996) Wooden ship building and the interpretation of shipwrecks. Texas A&M University Press, College Station
Stewart D (1999) Formation processes affecting submerged archaeological sites: an overview. Geoarchaeology 14(6):565–587
Torres R (2005) Estudo sobre a a distribuição e causalidade dos acidentes marítimos no litoral do Rio Grande do Sul, Brasil. Monography, Departamento de Oceanologia, Fundação Universidade Federal do Rio Grande, Rio Grande, Brazil
Torres R (2015) The archaeology of shore stranded shipwrecks of Southern Brazil. Unpublished PhD dissertation, Nautical Archaeology Program, Anthropology Department, Texas A&M University, College Station
Torres R (2017) Arqueologia náutica no sul do Brasil: Escavação dos sítios NAV Inédito e NAV Lagoa do Peixe. Cadernos do Lepaarq 14(27):333–365
UNESCO (1972) Underwater archaeology. A nascent discipline. UNESCO, Paris
UNESCO (1979) Conservación y Mejora de Playas URU.73.007. Ministerio de Transporte y Obras Públicas, Montevideo
Ward I, Larcombe P (2002) A process-orientated approach to archaeological site formation: application to semi-arid Northern Australia. J Archaeol Sci 30:1223–1236
Ward I, Larcombe P, Veth P (1999) A new process-based model for wreck site formation. J Archaeol Sci 26(5):561–570
Westerdahl C (1992) The maritime cultural landscape. Int J Naut Archaeol 21(1):5–14
Wheeler A (2002) Environmental controls on shipwreck preservation: the Irish context. J Archaeol Sci 29:1149–1159
Willey G, Philips P (1958) Method and theory in American archaeology. University of Chicago Press, Chicago
Williams A (1971) An analysis of some factors involved in the depth of disturbance of beach sand by waves. Mar Geol 11:145–158

Dunes as Units of Preservation of the Archaeological Record in the North Coast of San Matías Gulf, Province of Río Negro (Argentina)

Dalila Viviana Herrera Villegas

Abstract On the northern coast of the San Matías Gulf, dunes form the main units of the archaeological record; however, the dynamics of the coastal environments have caused many properties of this record to be affected, generating new spatial configurations. Therefore, the geoarchaeological perspective is a determining approach when dealing with different archaeological problems and contexts since it allows to know the diverse processes that acted in these contexts. The following paper presents an analysis of the Creek Bay archaeological locality. The main archaeological evidence is a series of shell middens located in dunes, which, due to their deposition dynamics, have resulted in a high degree of preservation. The objective of this paper is to identify the agents that acted on the preservation and current disposition of the archaeological record, recognizing the importance that the bodies of dunes have had in the state of conservation of the evidence, which is conformed by valves, remains of fish and marine mammals, terrestrial vertebrates, lithics, and carbon, among others.

Keywords Shell middens · Dunes · Geoarchaeology · Preservation

Introduction

On the Patagonian Atlantic coast, coastal dynamics over time have played a key role in the current configuration of the archaeological record, since they have made cultural evidence to acquire particular spatial and temporal properties. This phenomenon is observed in the northern coast of the San Matías Gulf (Province of Río Negro, Patagonia, Argentina) (Fig. 1) where it is common to find anthropic accumulations of valves (shell middens) showing differential degrees of preservation. The following

D. V. H. Villegas (✉)
INCUAPA – CONICET, Facultad de Ciencias Sociales, Departamento de Arqueología, Universidad Nacional del Centro de la Provincia de Buenos Aires (UNICEN), Av. Del Valle 5737, 7400 Olavarría, Argentina
e-mail: dalila1083@gmail.com

H. Inda Ferrero and F. García Rodríguez (eds.), *Advances in Coastal Geoarchaeology in Latin America*, The Latin American Studies Book Series, https://doi.org/10.1007/978-3-030-17828-4_6

Fig. 1 Location of the Creek Bay archaeological locality

paper seeks to identify the agents that have acted on the preservation and current disposition of the archaeological record.

The San Matías Gulf covers an area of 380 km^2 and is characterized by a semiarid climate, with an average temperature of 15 °C and an annual precipitation of less than 300 mm. The predominant winds come from the SW quadrant, and the vegetation is a bush type, corresponding to the Patagonian Monte formation (Cabrera and Willink 1980). The tides have a semidiurnal regime and are macromareal, with average amplitudes of 6.38 m and a maximum of 9.22 m in the movement of the San Antonio Este Port, in the San Antonio Bay (Servicio de Hidrografía Naval 2009).

The Creek Bay archaeological locality (Fig. 1) is characterized by a wide variety of archaeological and bioarchaeological remains that encompass the Middle and Late Holocene (Fig. 1). This locality belongs to one of the coastal entrances corresponding to topographic "bajos" that during the middle Holocene (ca. 7000 years BP) were flooded by the rise of sea level (Schellmann and Radtke 2010). Beginning with the middle Holocene transgressive maximum, this coast has undergone a major geomorphological evolution (Favier Dubois et al. 2016). The marine deposits consist mainly of offshore bars that are formed by terraces of varying elevations. During the Pleistocene, terrace elevations were $\geq$10 m above sea level. Holocene terraces are always below this level (Schellmann and Radtke 2010; Favier Dubois et al. 2016).

In this locality, different morphological groups of dunes are reported with dissimilar chronologies. The dunes origin is linked to the different pulses in sea level drop that occurred since the middle Holocene. On the northern coast of the gulf, the presence of Barchanoid dunes predominates, which stand out not only for being the most abundant but also due to their high dynamics of sedimentary deposition that facilitated the preservation of the archaeological evidence which also offers the best account of past human activity.

The archaeological record has been presenting a series of characteristics that have to do mainly with the generation of erosion foci in the dunes that have caused the archaeological record to be continuously exposed, redeposited, and displaced by wind, which also affects the qualitative and quantitative properties of the record and generates difficulties when making interpretations of the spatial and temporal contexts.

Materials and Methods

The methodology implemented in this paper consisted of a detailed analysis of the geological, archaeological, and paleoenvironmental literature, as well as the analysis of satellite images in which it was sought to identify the types of dunes present in Creek Bay.

In 2016, transects were undertaken in the locality that included different geomorphological units such as the current beach, the Holocene terrace, the dunes, and the shell middens in the dunes. In these spaces, sediment samples were taken for grain size analysis in order to determine the origin of the deposits and the sedimentary dynamics between the dunes and the beach, and if this remains represent the morphology of the dunes and the preservation of the cultural evidence. The structure of the sand grains and the mineral prevalence of each sample were determined by observing 10-g samples of in a stereoscopic microscope (Hokenn Optik model ZTX E ZOOM with 10× magnification).

For the analysis of particle size classification, USA standard ASTM E 11-61 was employed using sieves with different openings according to the grain size in each sample. For weighing the samples, an electronic balance (A&D FX 3000i) was used. The weight of the different fractions was recorded to create a histogram showing the predominance of the sedimentary particles in each of the collected samples. These results allow to determine how significant the beach–dune relationship remains, and its importance in the longer term preservation of deposits that have been uncovered, and of those that are just beginning their erosion process.

Chronological and Paleoclimatic Indicators

As previously mentioned, along this coastal strip, morphological groups of dunes with dissimilar chronologies are reported. These dunes are mainly parabolic and barjanoid. These are the product of variations in sand content related to changes in sea level. Geological and geomorphological studies conducted in the vicinity of the region account for arid conditions during the maximum transgressive of the Middle Holocene (ca. 7400 years BP), a period in which high global temperatures caused the sea level to increase between 6 and 7 m above the current level. Subsequently, toward 6000 years BP, there is a first pulse of descent of the marine level along the Patagonian

Coast (Schellmann and Radtke 2010) which is followed by a stage of stability until about 3000–2600 years BP. During this stage, when the beach sediment supply is not abundant because of the higher sea level, what it is seen today as fields of parabolic dunes began to develop. Then, between ca. 2600 and 2400 years BP, a second pulse, less than 1 m, of sea level drop was reported (Schellmann and Radtke 2010), which favored the development of aeolian bodies in the form of large Barjanoid dunes fields. These dune fields are directly related to a greater provision and mobility of sand, a condition that has made these dunes the most abundant along the entire northern coast of the gulf.

The paleoclimatic scenario of the Middle Holocene in this region corresponds to arid conditions along the northern Patagonian Coast (Schäbitz 1994, 2003; Favier Dubois et al. 2016). In this respect, the previous paleoclimatic studies carried out thus far consisted of pollen analyses in dry lagoons of northeastern Patagonia (Schäbitz 1994, 2003). These studies indicate arid conditions during the Middle Holocene (ca. 7500 years ^{14}C BP), with higher temperatures and lower rainfall levels. These arid conditions changed slightly to semiarid conditions between ca. 5000 ^{14}C years BP and ca. 3000 ^{14}C years BP. According to Schäbitz (1994, 2003), in ca. 3000–2500 ^{14}C years BP, semiarid conditions were established, with increasing annual rainfall levels (Schäbitz 2003).

Paleoenvironmental studies carried out on pollen obtained from two shovel tests at the Bajo de la Quinta Lagoon indicated the presence of a xerophytic shrub steppe under predominantly arid climatic conditions between 6800 and 3000 years ^{14}C BP, which then changes to semiarid conditions toward the late Holocene (ca. 3000 years ^{14}C BP). This change to semiarid conditions corresponds to the same scheme proposed by Schäbitz in the dry lagoons of northeastern Patagonia (Favier Dubois et al. 2016). However, around 1000 ^{14}C years BP, a new arid pulse is inferred by the reappearance of pollen types similar to the ones in the first-time block (Marcos et al. 2012). This pulse could represent anomalies of precipitation and temperature similar to the Medieval Warm Period, which was previously detected by dendroclimatological studies in NW Patagonia (Villalba 1994).

Geomorphological Description of Creek Bay Locality

On the northern coast of the San Matías Gulf, there is an important geomorphological feature formed by an enormous dune field that developed between the Caleta de los Loros and Creek Bay. This dune field covers up to 45 km inland with SW–NE direction (Fig. 2). This body of dunes evolved in an EW direction due to the intensity of the SW winds blowing from continental Patagonia (Favier Dubois and Borrella 2011).

The dunes in the Creek Bay locality are characterized as being mainly composed of a group of Barjanoid dunes, whose morphology is related to greater low sea levels during the late Holocene. However, in order for these bodies of sand to have a favorable development, vegetation must be present to provide cohesion and stability

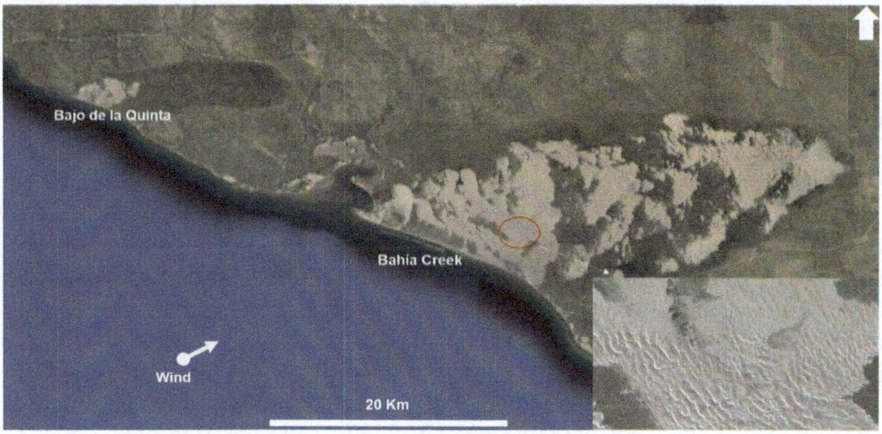

Fig. 2 Field of Barjanoid dunes which are highlighted in the box

of the surface, preventing the dunes from moving toward the interior of the continent. The vegetation also helps to prevent the dunes from being reactivated and displaced by the action of the wind. In order for this to happen, the vegetation that develops there must have a tolerance to burial, be resistant to heat, endure short droughts, and the limitation of nutrients (Tsoar 1990; Arens 1996).

Aeolian Morphology

In order for the dunes to acquire their characteristic morphology, a series of factors must be combined, among which are the constant supply of sedimentary material to the beach either by coastal drift or by sea currents and waves (Sanjaume et al. 2011), the presence of strong winds that come from the sea to the interior of the continent, moderate rainfall, minimal slope of the beach, areas with wide tidal range, low atmospheric humidity, and the presence of vegetation (McKee 1983).

Sedimentation Process in Dunes

To generate effective dune sedimentation and a characteristic morphology, a series of factors such as removal, transport, and sedimentation must be conjugated. Removal refers to the ability of the wind to lift sand particles (Warren 1979), which once having been agitated, can be moved by different mechanisms (creep, saltation, and suspension). After the particles are removed, the wind transportation processes vary in relation to the wind speed (Sanjaume et al. 2011). Finally, there is sedimentation, which is responsible for the balance and morphology of the dune, since there is a direct relationship between the wind and the creation of the dune. Sedimentation of the dunes occurs when the air is saturated with sand (Sanjaume et al. 2011) and it collides with an object, being it either vegetation or some other obstacle, which causes the intensity of the wind to decrease, leaving the sandy material trapped on the surface of the ground (Ley et al. 2007).

In the sedimentation process, vegetation plays an important role in the growth of coastal dunes, since it not only introduces roughness to the surface causing the wind speed to decrease but also favors the sedimentary development (Sanjaume et al. 2011). In this sense, when the first Psammophytic and Xerophytic communities, adapted to the harsh conditions of the dunes, were organized in strips, thus achieving stabilization as they moved away from the sea. However, when the vegetation disappeared, sedimentation decreased, which caused the sedimentary volume to be affected, changes that undoubtedly caused the gradual advance of the dunes, but also the development and morphology of the sedimentary structure (Ley et al. 2007).

Archaeological Background

To give an account of the importance of dune environments on archaeological sites, and with the purpose of not advancing much on the discussion, only a couple of Latin American works would be cited, which show the impact of different agents in the state of preservation of the archaeological record in such contexts, specifically regarding the incidence of such agents on the spatial and chronological arrangement of the archaeological record. The first work was published in Chile in 2001 and presented an investigation oriented to characterize and understand the processes of transformation and alteration of the archaeological materials found at the Punta Ñagué site. The studies carried out allowed to establish, on one hand, that the main processes of alteration and transformation that some settlements recorded are the product of the incidence of climatic factors and, on the other hand, they are the product of anthropogenic activity. Among the most outstanding changes, the author identified extremely fragmented mollusks, and the remains of marine mammals and birds presented discoloration, cracking, fracturing, and abrasion, among other consequences (Quintana 2001).

In 2015, a work was published in which the various factors that tend to impact archaeological sites in dunes are exposed; one of its objectives is to understand how the dune dynamics influenced the visibility of the archaeological record in the site RS-LC:73 located in the town of San José del Norte, Brazil. This study showed a variety of surface archaeological material that has undergone different cycles of erosion and deposition, highlighting that the bones and shells are the most affected by their exposure to weathering and erosion caused by the wind transport, also emphasizing the circulation of people and animals as disturbing agents, thus accelerating the degradation of the dunes, resulting in the elimination, displacement, and abrasion of cultural evidence (Silva Da Silveira and Serpa 2015).

For the specific case of Argentina, the archaeological works that have been carried out in different regions highlighted the importance of the vegetation on the dunes (Cruz et al. 2016; Borrero 2016; Borrazzo and Borrero 2016). First, the plants slowed the advance of the wind bodies, and second, they prevented them from being eroded, thus allowing the preservation of the archaeological material. The disappearance of the vegetation in the area, mainly due to anthropogenic causes, had consequences on the initial configuration of the cultural deposits (Cruz et al. 2016; Borrero 2016; Borrazzo and Borrero 2016). In response to the sedimentary dynamics, the archaeological record usually presented different degrees of preservation that are related mainly with two types of morphological modifications; the first one is linked with wind abrasion (corrasion), weathering and precipitation on the materials, and the second is given by spatial alterations of the archaeological evidence (displacements, transport, selective removal of materials, and deflation of the sedimentary matrix; Borrero 2016).

Results

The Archaeological Record at Creek Bay

The Creek Bay archaeological locality contained abundant evidence of hunter-gatherer occupations, since many of the sites still have good preservation. So far, human occupations have been dated between ca. 5310 and 1100 ^{14}C years BP (Favier Dubois and Borella 2007, Favier Dubois and Scartascini 2012). The archaeological sites are mainly represented by shellfish, mussels (*Mytilus edulis*), and "cholgas" (*Aulacomya atra*). There are also fish, marine mammals, terrestrial vertebrate remains (*Lama guanicoe*, armadillo, etc.), lithic materials, charcoal debris, etc. The shell middens are located on dunes that extend over the Pleistocene and Holocene marine terraces, located at varying distances from the current beach (Favier Dubois et al. 2016).

The Coastal Dunes and Their Relationship with the Properties of the Archaeological Record in Creek Bay

The beach–dune relationship is a determining factor not only in the morphology of the dune, but in the preservation of the contextual properties of the archaeological record. To evaluate the relationship between the beach and the dunes, a sampling transect was carried out in different geomorphological units that include the current beach, the Holocene terrace, a dune on the Holocene terrace, and a shell midden located on a dune (Fig. 3).

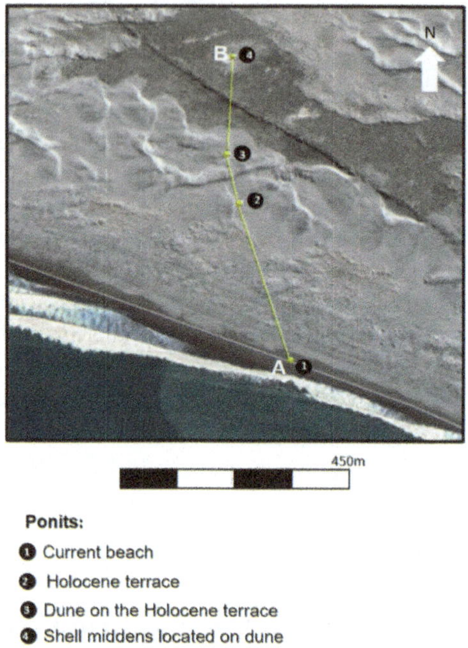

Fig. 3 The drawn line shows the transect carried out in the geomorphological units of Bahía Creek; in the box, the arrow shows the location where the sedimentary samples were collected

Grain Size Results

The results obtained from the weight and the percentage of each sample particle size showed that the mean grain size is between 0.297 and 0.149 mm. The morphology of the grains oscillates between subangular and subrounded, with predominance of quartz and feldspar minerals.

With these results, the grain size distribution of all the samples was identified between fine and very fine intervals. A predominance of fine sand (Fig. 4) is in agreement with its aeolian origin, that is, with the ability of the wind to remove, transport, and deposit the mineral particles, confirming in this way the close relationship that still exists between the beach as the source material for the dunes.

Due to the constant and strong winds from the western quadrant, the sedimentary dynamics increased and with it the probabilities that the morphology and mobility of the dunes will remain active. This in turn favors the rapid burial and the preservation of the archaeological deposits. Subsequently, the vegetation stabilized the sand bodies that were, then, away from the areas of contribution as the coast of the bay was filled by offshore bars.

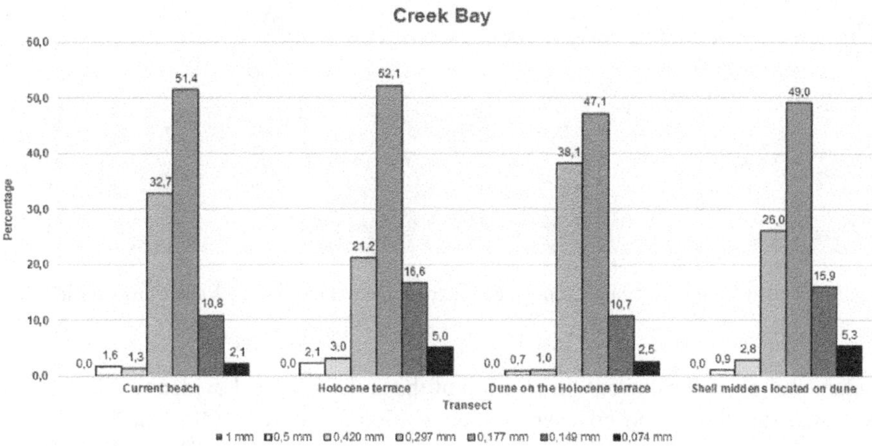

Fig. 4 Histogram of grain size percentages of each geomorphological unit

Degradation of the Dunes and Their Effect on the Properties of the Archaeological Record in Bay Creek

As a result of the archaeological and geoarchaeological studies carried out from 2004 to 2010 by Cristian Favier Dubois and his research team, it was possible to verify by means of excavations and surface surveys that the materials from the shell middens did not remain exposed for long periods of time. The material remains showed no signs of abrasion, solar calcination, weathering, fragmentation, or dispersion (Favier Dubois et al. 2016).

Recent field work showed that the dunes which contained archaeological deposits were affected by different factors, such as the presence of livestock (sheep and cattle), modern day walking paths, roads, and recreational activities such as of 4 × 4 trails (Favier Dubois and Borrella 2007). Consequently, these processes eroded and modified the native vegetation cover. Subsequent to the loss of vegetation, this resulted in the degradation of the dunes. Therefore, archaeological materials began to become exposed, and then removed, displaced, and redeposited (Fig. 5), an aspect that clearly affects the understanding of the archaeological context and its properties.

Modification of the Spatiotemporal, Qualitative, and Quantitative Properties of the Archaeological Record at Creek Bay Locality

As a result of the environmental and anthropogenic factors that have acted over time in Creek Bay, it is possible to identify a series of factors that contribute to the

Fig. 5 Deposit features. **a** tabular deposit geometry, **b** mound morphology, and **c** eroded shells lens

understanding of the current disposition of the archaeological evidence. The aeolian processes that have exposed a series of depositional features such as the topography, geometry, stratigraphy, and morphology of the shell middens (Favier Dubois and Borella 2007; Favier Dubois et al. 2016) provide important information about the history of formation of the archaeological record. Yet when these characteristics are exposed, the thickness of the accumulations is threatened, which increases the risk of the original lenses being eroded (Fig. 5), preventing a correct interpretation of the deposits (Favier Dubois and Borella 2007). In the best of circumstances, these data will be relevant for understanding the differences of preservation between the materials and their spatiotemporal qualities.

The spatial relationships that may exist between the different elements that make up the archaeological record can be substantially modified by superficial aeolian processes (Butzer 1982; Waters 1992 and Schiffer 1987), which causes the original systemic context of a site to be modified before or after its burial. In this sense, in the locality under study it has become a constant to find remains of shells exposed on the surface, a condition that has sufficed to cause changes in the morphology of the terrain since these, when exposed, can be displaced by the wind as well as gravity, resulting in a floor extension ranging in variation between two and six meters (Favier Dubois et al. 2016) (Fig. 6). These changes cause an increase in deflation in the peripheral sectors of the shell middens, making the length and angle of the gradient of the bank slope larger, and thus favoring the gravitational drag of the cultural remains (Quintana 2001).

When archaeological materials are exposed to the external environment, not only do they suffer differential weathering, but they are also fragmented and transported at varying distances. Weathering and fragmentation cause the materials to be easily displaced by aeolian action toward the interdune zones, where the material is most likely mixed with different deposits (Schiffer 1983; Waters 1992), forming a series of "spatial palimpsests" (Bailey 2007). In some cases, this may even give way to what the present author has defined as "false spaces of human occupation" (Fig. 6). These spaces do not necessarily correspond to specific areas of human activities, although it is very likely that they will be assigned to this category if the particularities of each depositional environment are not taken into account.

Fig. 6 **a** Deflated shell midden, **b** morphology derived from aeolian erosion

From the point of view of the conservation of shell middens, it is still possible to demonstrate a good resolution and integrity. Resolution is understood as the number of events represented in a deposit, and for integrity, the number of natural and anthropic agents that are involved in the process of formation of the archaeological record (Binford 1981). With this in mind, it is fundamental to consider that the composition of the archaeological record will not necessarily consist exclusively of evidence of past human activity, because, regardless of the context of deposition, much of the archaeological record has been formed by the same processes that have modeled the landscape (Butzer 1982; Schiffer 1987; Waters 1992). This leads to consider that these geodynamic processes will have to be represented in some manner on the contextual properties of cultural evidence.

The archaeological material recovered during the 2004, 2010, and 2016 fieldwork seasons presented evidence of taphonomic modifications such as abrasion and/or corrosion, which suggested that the materials have experienced a number of exposure and burial events generated mainly by strong winds and sand. Thus, the main cause of the exposure, fragmentation, and displacement of the different materials is due to the modern anthropogenic activities that have taken place in this locality.

The main taphonomic modifications identified in the bone remains of land birds and marine mammals include discoloration, fracturing, cracking, and exfoliation. In addition, root etching suggests that at some point the dunes contained plant communities. Similarly, in many sectors of Creek Bay, the vegetation cover was eliminated, so that until now, the only thing that indicates moments of greater stability has been the presence of these root marks (rhizocrecretions). Likewise, the rhizoconcretions suggest a period of environmental stability, in which erosion and sedimentation decreased. These characteristics are crucial when considering the pedogenetic development since they are directly linked to the presence of a vegetation cover and the stabilization of sandy structures.

Although to date no pedological features have been observed in any excavated site, except for rhizoconcretions (Favier Dubois and Borella 2007), and it should be noted that these develop in subsurface soil horizons (usually the C horizon), indicating that the A horizon has been eroded prior to the exposure of the shell midden (Favier Dubois and Borella 2007).

Conclusions

The information presented in this paper suggests that the preservation of archaeological remains from the Creek Bay locality is the result of a second pulse of marine regression during the Late Holocene (2600–2400 years BP). During this period, the sedimentation was higher, which favored the formation of Barjanoid dunes and the rapid burial of archaeological deposits.

In spite of diverse erosive agents that acted on the shell middens, modern anthropogenic activity is one of the main disturbing agents and thus primarily responsible for the degradation of cultural remains. A large quantity and variety of the archaeological material are found on deflated surfaces and in the interdune areas (Manzi et al. 2011), which, in spite of the previously mentioned conditions, still present a state of differential preservation that has to do with the different cycles of erosion and burial to which they have been subjected.

Based on the grain size results made on the sediment samples collected in the geomorphological units, the differential preservation in many of these sites is likely due to the constant input of sedimentary material coming from the beach, an issue that to date has allowed the preservation in some cases of the original characteristics of archaeological deposits.

The optimal conditions of sedimentation that still persist in this locality offer, on one hand, the preservation of the shell middens that have just begun their exposure, and on the other, they would improve the preservation conditions of the deposits that have already been exposed due to eolian processes.

The geo-environmental dynamics have proved to have an important role in the modeling of the properties of the coastal archaeological record. The geoarchaeology analysis performed here seeks to account for the factors that affected not only the qualitative and quantitative particularities of the archaeological record, but also the spatial and temporal modifications caused by natural processes. These processes can be misinterpreted as a cultural pattern, leading to incorrect interpretations about the use and function of spaces by past human populations.

Acknowledgements Funding for this project was provided by a grant from the Agencia Nacional de Promoción Científica y Tecnológica of Argentina, PICT-1128. The author would like to thank Dr. Cristian Favier Dubois for his support and valuable suggestions. Veronica Lalin de Aguilar and Daniel Rafuse helped with the translation, corrections, and suggestions to improve the manuscript. To the organizers of the IV workshop of the Study Group of Geoarchaeology of Latin America GEGAL for the management and invitation to participate in this publication. To the Facultad de Ingeniería of UNICEN, for providing spaces and equipment.

References

Arens S (1996) Patterns of sand transport on vegetated foredunes. Geomorphology 17:339–350

Bailey G (2007) Time perspectives, palimpsests and the archaeology of time. J Anthropol Archaeol 26(2):198–223

Binford LR (1981) Bones: ancient men and modern myths. Academic Press, New York

Borrazzo K, Borrero L (2016) Caracterización del modo tafonómico dunas en Fuego Patagonia: Aportes de la localidad Chorrillos (Tierra Del Fuego, Argentina). In: Actas del XIX Congreso Nacional de Arqueología Argentina. Facultad de Ciencias Naturales and Instituto Miguel Lillo, Universidad Nacional de Tucumán, Argentina. Serie Monográfica y Didáctica, vol 54. San Miguel de Tucumán

Borrero L (2016) Tafonomía y procesos de formación del registro arqueológico en ambiente de dunas: hacia la definición de un modo tafonómico. In: Actas del XIX Congreso Nacional de Arqueología Argentina. Facultad de Ciencias Naturales and Instituto Miguel Lillo, Universidad Nacional de Tucumán, Argentina. Serie Monográfica y Didáctica, vol 54. San Miguel de Tucumán

Butzer KW (1982) Arqueología, una ecología del hombre: Método y teoría para un enfoque contextual. Ediciones Bellaterra, Barcelona

Cabrera AL, Willink A (1980) Biogeografía de América Latina. OEA Monographs (Washington D.C.)

Cruz I, Ercolano V, Muñoz S, Lemaire C (2016) Procesos de formación del registro arqueológico en dunas costeras: el caso de Punta Entrada (Santa Cruz, Argentina). In: Actas del XIX Congreso Nacional de Arqueología Argentina. Facultad de Ciencias Naturales and Instituto Miguel Lillo, Universidad Nacional de Tucumán, Argentina. Serie Monográfica y Didáctica, vol 54, San Miguel de Tucumán

Favier Dubois C, Borella F (2007) Consideraciones acerca de los procesos de formación de concheros en la costa norte del Golfo San Matías (Rio Negro, Argentina). CONICET – INCUAPA, Facultad de Ciencias Sociales (UNCPBA), Olavarría

Favier Dubois CM, Borella F (2011) Contrastes en la costa del golfo: una aproximación al estudio del uso humano del litoral rionegrino en el pasado. In: Arqueología de pescadores y marisqueadores en Nordpatagonia. Descifrando un registro de más de 6000 años. Compiled by F. Borella and M. Cardillo, pp 13–42. Dunken, Buenos Aires

Favier Dubois CM, Scartascini F (2012) Intensive fishery scenarios on the North Patagonian coast (Río Negro, Argentina) during the Mid-Holocene. Quatern Int 256(1):62–70

Favier Dubois C, Kokot R, Scartascini F, Borella F (2016) Una perspectiva geoarqueológica del registro de ocupaciones humanas en el golfo San Matías (Río Negro, Argentina). In: Intersecciones en Antropología, special volume 4, pp 47–59. La geoarqueología en la Argentina: Aportes y avances. Facultad de Ciencias Sociales, Universidad Nacional del Centro de la Provincia de Buenos Aires, Argentina

Ley C, Gallego JB, Vidal C (2007) Manual de Restauración de dunas costeras. Dirección General de Costas, Ministerio del Medio Ambiente, Madrid, p 248

Manzi L, Borella F, Cardillo M (2011) Distribuciones artefactuales: Una aproximación a la estructura espacial del registro arqueológico del litoral atlántico rionegrino. In: Arqueología de Pescadores y Marisqueadores en Nordpatagonia. Descifrando un registro de más de 6.000 años. Editorial Dunken, Buenos Aires, pp 44–66

Marcos MA, Mancini M, Favier Dubois CM (2012) Middle-Late Holocene environmental changes in Bajo de la Quinta, NE Patagonia, inferred by palynological records and their relation to human occupation. The Holocene 256(11):1271–1281

McKee ED (1983) Eolian sand bodies of the world. In: Brookfield ME, Ahlbrandt TS (eds) Eolian sediments and proceses. Elsevier, Amsterdam, pp 1–26

Quintana R (2001) El sitio arqueológico Punta Ñagué: Procesos de transformación y estrategias de conservación durante su excavación. En Conserva, n°5

Sanjaume E, Gracia FJ, Flor G (2011) Introducción a la Geomorfología de sistemas dunares. In: Sanjaume ES, García PJ (eds) Las dunas en España. Sociedad Española de Geomorfología, pp 10–32. Universidad Gallega, Spain
Schäbitz F (2003) Estudios polínicos del Cuaternario en las regiones áridas del sur de Argentina. Revista del Museo Argentino de Ciencias Naturales 5(2):291–299. Buenos Aires
Schäbitz F (1994) Holocene climatic variations in Northern Patagonia, Argentina. Palaeogeogr Palaeoclimatol Palaeoecol 109:287–294. (2003) and Estudios polínicos del Cuaternario en las regiones áridas del sur de Argentina. Revista del Museo Argentino de Ciencias Naturales 5(2):291–299. Buenos Aires
Schellmann G, Radtke U (2010) Timing and magnitude of Holocene sea-level changes along the middle and south Patagonian Atlantic coast derived from beach ridge systems, littoral terraces and valley-mouth terraces. Earth Sci Rev 103:1–30
Schiffer MB (1983) Toward the identification of formation processes. Am Antiq J Soc Am Archaeol 48(4):673–697
Schiffer MB (1987) Formation processes of the archaeological record. University of New México Press, Albuquerque, USA
Servicio de Hidrografía Naval (2009) Tablas de Marea. Servicio de Hidrografía Naval, Armada Argentina, 643 pp. Buenos Aires
Silva da Silveira C, Serpa C (2015) Visibilidade de Vestígios Arqueológicos e sua Relação com a Morfodinâmica Dunar, Estudo de caso do sítio RS-LC: 73. In: Cadernos do LEPAARQ, vol XII, n°23
Tsoar H (1990) Trends in the development of sand dunes along the Southeastern Mediterranean coast. In: Bakker TW, Jungerius PD, Klijn JA (eds) Dunes of the European coasts. Catena Supp. 18, pp 51–60
Villalba R (1994) Tree ring and glacial evidence for the medieval warm epoch and the little ice age in Southern South America. Clim Change 26:183–197
Warren A (1979) Eolian processes. In: Emblenton C, Thornes J (eds) Processes in geomorphology. Arnold, London, pp 325–351
Waters MR (1992) Principles of geoarchaeology. The University of Arizona Press, Tucson & London

Mounds and Landscape in the Merín Lagoon Basin, Uruguay

Roberto Bracco Boksar, Daniel Panario, Ofelia Gutiérrez, Andreina Bazzino, Christopher Duarte, Rosario Odino and Ethel Reina

Abstract In this work, a series of data is presented, which is obtained by elemental analysis and edaphic analysis of the matrix of three mounds of the García Ricci site, located to the south of the basin of the Merín lagoon, Uruguay. These data will contribute in increasing the knowledge of the construction processes of the mounds. The elemental composition data of the matrix indicate possible variations in the composition and origin of the materials used to elevate the mounds. The data obtained by edaphological techniques would indicate that during the construction, significant amounts of organic matter were incorporated. But the relatively low amounts of organic matter in the matrix lead us to admit that it had to be degraded. By focusing on the large amounts of burned earth that compose the Matrix, we were able to consider that part of the organic matter could have been plant fuel. This would have burned to promote the thermal alteration of the sediments that form the mounds. This paper refers to the traditional practices of soil thermal alteration to address the possible procedures and their consequences. Finally, it is pointed out that the recur-

R. Bracco Boksar
Facultad de Humanidades y Ciencias de la Educación, Universidad de la República, Av. Uruguay 1695, 11200 Montevideo, Uruguay

Laboratorio 14C, Facultad de Química, Universidad de la República - Ministerio de Educación y Cultura, General Flores 2124, 11200 Montevideo, Uruguay

CURE, Sede Rocha, Universidad de la República, Route No. 9 Km 207, 27000 Rocha, Uruguay

D. Panario · O. Gutiérrez (✉)
UNCIEP, Instituto de Ecología y Ciencias Ambientales (IECA), Facultad de Ciencias, Universidad de la República, Igua 4225, 11400 Montevideo, Uruguay
e-mail: oguti@fcien.edu.uy

A. Bazzino
Equipo técnico Proyecto RLA0058-AIEA, Montevideo, Uruguay

C. Duarte
Laboratorio OSL, UNCIEP, Instituto de Ecología y Ciencias Ambientales (IECA), Facultad de Ciencias, Universidad de la República, Iguá 4225, 11400 Montevideo, Uruguay

R. Odino · E. Reina
Dirección Nacional de Minería y Geología (DINAMIGE), MIEM, Hervidero 2861, 11800 Montevideo, Uruguay

H. Inda Ferrero and F. García Rodríguez (eds.), *Advances in Coastal Geoarchaeology in Latin America*, The Latin American Studies Book Series, https://doi.org/10.1007/978-3-030-17828-4_7

sive use of these practices led to the formation of patches within the landscape. Its implantation and growth for millennia were gradually modifying the landscape and particularly the distribution and circulation of plant and animal species, including humans. Pyrotechnology would have played a very important role in this modification.

Keywords Mounds · Merín Lagoon basin · Prehistoric landscapes · Mound geochemistry · Pyrotechnology

Introduction

Since the mid-1980s, Uruguayan archaeology has focused on the mounds of the Merin Lagoon basin, eastern Uruguay. Between 5.5 and 0.2 ka ^{14}C thousands of mounds ("cerritos de indios", "aterros") were built in the basin of the Merin Lagoon and adjacent areas, in flooding lowlands ecosystems (locally named as "bañados" and "esteros"), or in its surroundings (see Schmitz 1976, 1981; Copé 1991; Cabrera 1999, 2013; Bracco et al. 2000c; Gianotti 2000; Iriarte et al. 2000; López 2001; López et al. 2001; Iriarte 2003; Bracco 2006; Milheira et al. 2016) (Fig. 1).

These mounds have a variable height of some decimeters to more than 7 m and a circular-to-subcircular base with a diameter of approximately 35 m (Fig. 2). They have been characterized as earth constructions that integrate the waste of human activities (bone remains, lithic artifacts, and pottery sherds) as well as human burials. In the study of its genesis, attention has been paid mainly to the collection, transport, and deposition of sediments and discarded elements (Schmitz 1976, 1981; Bracco et al. 2000b; Castiñeira and Piñeiro 2000; Iriarte et al. 2000; López 2001; Cabrera 2013; Milheira et al. 2016).

The mounds are located in two different landscape units: low and middle plains ("bañados" and "esteros") and gentle hills and rocky hills ("lomadas" and "sierras"). The first unit is the domain of the flat geoforms, with a deep sedimentary development, medium textures, slow drainage, with intermittent hydromorphism (Panario et al. 2015); it integrates a series of lowlands that flood temporarily or permanently (PROBIDES 1999: 55). Here, the mounds follow the braided watercourses of the wetlands (Bracco et al. 2015). The second unit is characterized by a landscape of soft to strongly undulating reliefs with a matrix of summer meadows. Here, the mounds are found mainly at the end of the soft hills that project towards or are surrounded by plains.

In the lowland plains, mounds are the only positive relief features in the landscape. In the southern part of the basin, the largest concentration of mounds—including the highest and oldest ones—is located in the region of India Muerta–Paso Barranca (Bracco et al. 2015) (see Fig. 1). Its density reaches 0.6 mounds per sq. km. They can be found as isolated structures but, more frequently, they occur in groups. The most numerous groups are located in soft hills, reaching a population of 50 mounds in 0.4 km^2 (MD site, 34°41′21.52″S, and 53°56′34.45″W) (Bracco et al. 2000a, 2015).

Fig. 1 Map of the Merín Lagoon basin. The sectors where the sites with mounds are grouped (gray areas) are indicated. (1) García Ricci site, (2) PSL site, and (3) site CH2D01

The function of these structures has been subject to long controversies. It has been proposed that they were erected as platforms that allowed the occupation of wetlands during floods; they were also interpreted as funerary sites, ritual sites, territorial markers, or multifunctional systems for agricultural purposes (Schmitz 1976, 1981; Baeza and Panario 1999; Bracco et al. 2000a; Gianotti 2000; Iriarte et al. 2000; López 2001; Gianotti et al. 2008; Cabrera 2013; Milheira et al. 2016). The ^{14}C and OSL/TL dating sets have shown that the mounds were elevated for very long periods (from centuries to millennia), which has led to rethink many of the proposed functions (Bracco and Ures 1999, 2001; Bracco 2006; Santos et al. 2015).

From the geoarchaeological point of view, focusing on the relationship between the environment and human societies, this project has tried to identify the conditions and causes of the emergence and continuity of the behavior that led to the elevation and occupation of these mounds. More specifically, the regional paleolimnological, paleohydrographical, and paleoclimatic information have been integrated with the objective of establishing a paleoecological framework from the Middle Holocene to the present and correlating it with the variations of the archaeological record (Bracco et al. 2005, 2011; García-Rodríguez et al. 2010; del Puerto 2011, 2015; Inda 2011). Different proxies have pointed out that at the beginning of the construction of mounds—around 5500 yr ^{14}C BP—the climate changed towards drier and colder conditions, in comparison with the current ones (Bracco et al. 2011; del Puerto 2015). Based on this, the concentration of settlements in the lowlands was interpreted as an adaptive response to stressful climatic/environmental conditions. This would have originated settlements located in the areas with greater water availability and therefore other resources, converting the elevation of the mounds into a distinctive form of occupation of those spaces (Iriarte 2003; Bracco 2006; Bracco et al. 2011).

The archaeological record has provided evidence that the human groups that elevated the mounds were hunter-gatherer-fishers who integrated agriculture in a subsidiary way (Bracco et al. 2008; del Puerto 2015). However, Iriarte (2003) put more emphasis on regional agriculture. The archaeofaunal record is dominated by deer and fish, while the macro botanical record gave evidence of the use of fruits from local palm trees. Opal phytoliths indicated an early incorporation of cucurbits, beans, and corn around the 3000 yr ^{14}C BP (Iriarte 2003), although del Puerto (2015) argued that strong evidence of domesticated species appeared only after 2000 yr ^{14}C BP. However, the isotopic signal ($^{13}C_{col}$, $^{13}C_{ap}$, and ^{15}N) of human remains, showed that the mound builders based their economy on hunting-gathering and fishing, focusing on the naturally available resources in the inland sectors of the wetlands. Horticulture played a minor role, at least in terms of dietary input (Bracco et al. 2000b; del Puerto 2015; Mut 2015).

Fig. 2 Mound of the PSL site located in Paso Barranca–India Muerta locality

Based on the distributive and hierarchical analyses, the mounds have been reconceptualized as landscape ecology patches, restricted areas caused by a disturbance that differ markedly from the surrounding matrix (Forman and Godron 1986). In this case, the disturbance is the consequence of human activity and the contrast with the matrix is given by its relief shape, its edaphic characteristics, and its vegetation (see Fig. 2). The interaction of the mounds with the matrix determined the characteristics of these disturbances that, since their implantation (and growth), influenced the distribution and circulation of the biota (including those that built them).

In this work, the study of the mounds located in India Muerta–Paso Barranca (see Fig. 1) was approached, from a perspective of Landscape Ecology and Historical Ecology. Although both approaches are similar, from the first, the concept of landscape was taken (its components and specifically the concept of patch) and the temporal perspective from the second. Historical Ecology seeks to build an image of the relationships between man and the environment for a specific time and place, taking into account a wide range of biological, physical, ecological, social, and human tests (Terrell et al. 2003). It forms a different insight on the human–environment relation, rejecting from scratch the idea of "adaptation" and replacing it with "response". From this point of view, human beings not only live in their environment but also create, modify, and administer it thanks to the multigenerational accumulated knowledge that resides in their culture (Balée and Erickson 2006; Erickson 2006, 2008).

From this theoretical framework and observing that one of the dimensions that differentiates mounds as patches of their environment are the characteristics of their matrix, a line of research that focuses on their analysis was developed. The work of Browne Ribeiro (2011: 85) has been followed, who has pointed out the informative potential of the archaeological site matrix, paying special attention to phosphorus compounds as one of the geochemical records most clearly linked to ancient human activity (Holliday and Gartner 2007). In this work, the first results of the elemental chemical analysis and the edaphological analysis (phosphates, organic matter, carbonates, and pH) obtained from the García Ricci site matrix are communicated and interpreted. In addition, new luminescent dates are reported, which together with the available dates of OSL, TL, and ^{14}C, improve the chronological framework of the investigation.

Historical Background

Edaphology and Geochemistry

The first edaphic analysis of the matrix of the mounds was made in the 1990s decade (Durán 1990) for the CH2D01 site, located at the San Miguel wetland area (33°43′51.62″S/53°36′19.33″O) (see Fig. 1). It is composed of two mounds of approximately 1.2 m high (Curbelo et al. 1990). Two ^{14}C datings came from the

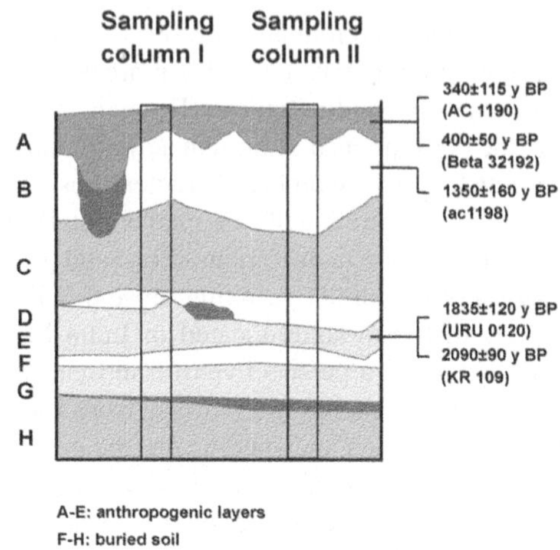

Fig. 3 Stratigraphy mound A site CH2D01. Sampling columns are indicated (taken from Bracco and Ures 2001, Fig. 1)

base of the mound “A”: 1835 ± 120 ^{14}C yr BP and 2090 ± 90 ^{14}C yr BP. Two others come from its top: 340 ± 115 ^{14}C yr BP and 460 ± 50 ^{14}C yr BP. The base and top from mound “B” were dated in 2430 ± 60 ^{14}C yr BP and 1090 ± 70 ^{14}C yr BP, respectively (Bracco and Ures 1999, 2001). For the matrix analysis, the northern face of the excavation I was sampled, which was located at the center of the mound “A”. In this profile, five main stratigraphic units were recognized perceptually as a product of anthropogenic accumulation (A–E, 0–128 cm depth) and three others (F–H) that corresponded to the underlying soil (Fig. 3). Samples were taken from the central part of each of the eight stratigraphic units, in two columns distant 1.5 m to each other, to observe possible horizontal variations. Table 1 shows the reported data of grain size, pH, organic material (OM), available phosphorus, and potassium (Durán 1990).[1] Figure 4 shows horizontal and vertical variations in the plotted data. The contrast between anthropogenic and underlying natural stratigraphic units is given mainly by textural differences, higher percentages of organic matter, and available phosphorus.

[1]Techniques used were: pH potentiometer, OM by Walkley and Black, available P by Bray, exchangeable potassium by flame photometry in extract with AcONH4 and granulometry by international method (Durán 1990: 17).

Table 1 Grain size data (sand, silt, clay), pH, organic material (OM), available phosphorus (P) and potassium (K^+) for the Mound "A" of the CH2D01 site (Durán 1990). From 0 to 128 cm deep (A–E), stratigraphic units product of anthropic accumulation, and 128–198 cm (F–H) correspond to buried soil below

Stratigraphic unit	Depth (cm)	Sand	Silt	Clay	pH	OM (%)	P (ppm)	K+ (cmol/kg)
		(%)						
Column I								
A	16.5	24.2	49.7	26.1	5.3	3.6	315	1.6
B	41.0	36.1	51.4	12.5	6.2	2.1	180	3.0
C	68.5	31.4	54.0	14.6	6.4	2.3	180	3.2
D	101.5	33.0	52.1	14.9	6.6	3.3	180	3.3
E	128.5	18.8	69.4	11.8	6.6	1.4	140	3.2
F	143.0	27.1	56.7	16.2	6.6	1.7	150	3.2
G	165.0	19.0	57.3	23.7	6.6	1.6	170	2.7
H	198.0	16.7	48.8	34.5	6.5	0.4	120	3.2
Column II								
A	16.5	27.6	48.6	23.8	5.6	5.0	330	1.8
B	41.0	34.0	50.0	16.0	6.2	1.6	330	3.0
C	68.5	34.2	47.6	18.2	6.5	2.0	275	2.8
D	101.5	23.0	63.0	14.0	6.6	1.9	205	3.3
E	128.5	23.9	59.2	16.9	6.6	1.9	175	3.2
F	143.0	19.4	54.2	26.4	6.5	1.9	195	2.5
G	165.0	21.5	54.4	24.1	6.3	1.2	215	2.4
H	198.0	13.8	47.4	38.8	6.1	0.3	135	3.5

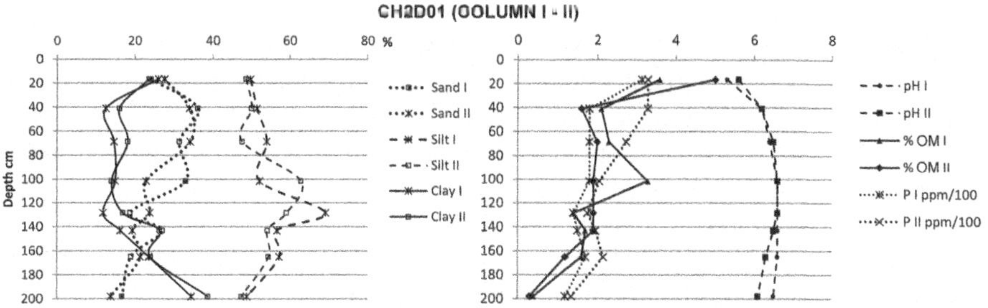

Fig. 4 Left: textural data from the two sampling columns of mound A, site CH2D01. Right: organic matter, pH and phosphorus available from the two sample columns of mound A, site CH2D01 (taken from Durán 1990)

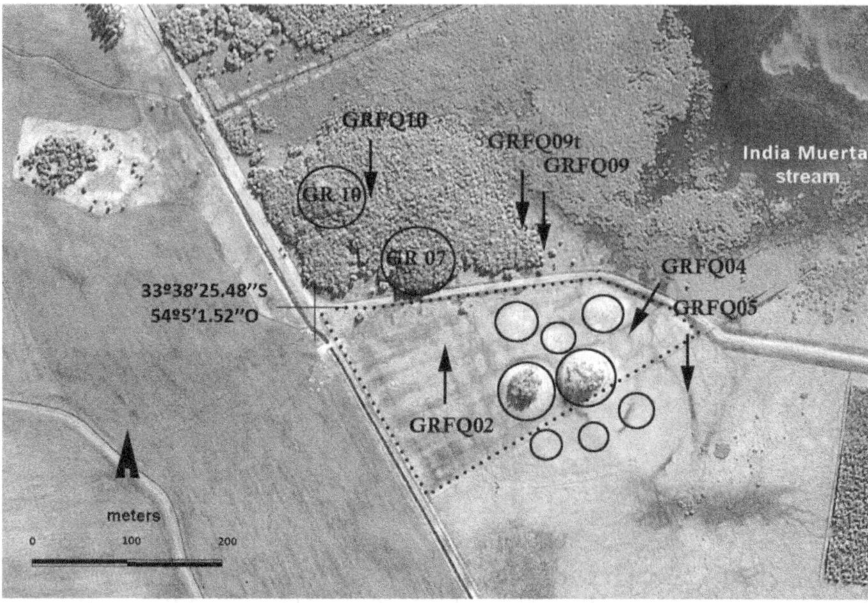

Fig. 5 Satellite image (Google Earth-2014) of the García Ricci site. The circles indicate the mounds. The arrows point the sampling points out of the mounds. The dotted area indicates the area that was cultivated in the 1970s–80s

Materials and Methods

Study Area

The García Ricci site (33°38.427′S–54°4.891′W) is located on the right bank of the India Muerta stream, in the center of the department of Rocha (Uruguay), in the region of India Muerta–Paso Barrancas, at the southern sector of the Merín Lagoon basin (see Fig. 3). It has 10 mounds, of which 2 of them exhibit a height of approximately 4 m, two others exhibit a height of approximately 3 m, and the rest are less than 1.5 m high (see Figs. 1 and 5).

The India Muerta–Paso Barrancas region is an extensive plain developed from deep sedimentary deposits, which are flooded permanently or temporarily (Panario et al. 2015). Its current climate is tempered with a subtropical tendency. The average annual temperature is 16 °C. The annual and daily thermal amplitude is moderated by the proximity of the ocean. The annual rainfall varies between 1,123 and 1,293 mm, without there being a rainy season (PROBIDES 1999). In this region, whose area is approximately 900 km^2, 174 sites were identified that accumulated more than 600 mounds. It is the area of plains in the southern sector of the Merín Lagoon basin with the highest concentration of mound sites and where the highest and oldest mounds are found (Bracco et al. 2015). The ^{14}C chronologies indicate that the majority were built between 4500 and 2000 yr ^{14}C BP (3209—6 Cal yr BC; see Reimer et al. 2013) although two have shown ages of around 5300 Cal. yr BC at their base (see Bracco et al. 2015: Table 1).

Sampling and Analysis

Samples were taken from the two highest mounds (mounds 7 and 10) of the García Ricci site and from a third one approximately 1.5 m high (mound 5). An AMS sampler was used, recovering from the top to the base, successive undisturbed vertical samples, jacked in PVC tubes of 15 cm in length. Those destined to be dated by TL/OSL were extracted by preserving them from exposure to light. In addition, samples were taken from the A horizon in 6 points of the surrounding plain and two anthills (see Fig. 5). Today, the density of these epigeal nests of *Camponotus punctulatus* around the mounds is of 400/ha.

The techniques used were elementary analysis by X-ray fluorescence (XRF), available phosphorus analysis (Bray and Kurtz 1945), organic matter, and carbonates determination by ignition (Heiri et al. 2001) and pH (H_2O). The XRF analyses were conducted at the National Direction of Mining and Geology of the Ministry of Industry, Energy and Mining, Uruguay. Soil analyses were carried out in the Soil Laboratory of the Faculty of Sciences of the Universidad de la República, Uruguay.

To select the vertical intervals between the analyzed samples of the mounds, the variations of the elementary composition of six sections of the column sampled for mound 7 were taken into account. These sections were subsampled every 2 or 4 cm (Table 2).

A set of aliquots of the samples taken at the plain and the epigeal nests were heated in muffle at 100, 200, 300, 400, 500, and 600 °C for 20 min, before measuring their levels of available phosphorus.

The chronology was constructed by TLD and OSLD. The analyses were made in the Luminescence Laboratory of the Faculty of Sciences, Universidad de la República, Uruguay. Two fractions of the mound matrix were dated: concretions of burned sediments and fine sediments (Duarte et al. 2017). The concretions of the burned sediments show evidence of thermal alteration on all their faces (color, hardness) and can reach more than 5 cm in diameter. Between 25 and 30% of the matrix of the levels that were dated were constituted by burned earth. Similar percentages were observed in the four mounds excavated from the PSL site, located 20 km away (Bracco et al. 2000c; Castiñeira et al. 2015).

Table 2 Results (ppm) from the elementary analysis (XRF) for the mound 7, site García Ricci

Depth (cm)	Al	Si	P	K	Ca	Ti	Mn	Fe	Sr	Zr	Ba
18	33,527	308,826	2086	15,240	10,370	6060	656	16,810	227.5	407.6	663
24	39,527	336,357	2095	16,220	12,380	5977	1079	16,640	237.6	440.7	654
26	40,793	330,239	2641	16,380	13,990	6000	1264	17,000	248.7	406.4	653
28	40,341	337,288	2718	16,250	14,890	6052	1063	16,520	249.8	412.9	619
30	41,833	347,263	2384	16,560	13,530	6053	966	16,590	249.1	422.7	644
32	43,166	347,396	2170	16,830	13,360	6279	1134	16,650	247.9	435.3	667
42	36,092	30,0048	1673	16,070	12,320	6421	832	18,420	239.0	427.8	644
54	34,431	293,265	1489	16,160	11,890	6153	692	18,300	239.2	417.7	596
62	36,804	296,590	1747	17,040	13,900	5958	777	18,440	260.1	390.0	635
90	34,736	311,087	1819	17,730	15,640	6154	820	17,240	275.0	388.0	642
98	40,228	345,401	2425	18,300	16,490	6170	808	16,510	276.4	399.1	596
100	41,008	343,140	2633	18,640	17,070	6052	846	16,960	274.6	418.4	604
102	39,527	338,884	2616	18,510	17,120	6224	813	16,480	278.6	443.3	612
104	40,680	352,583	2783	18,990	17,770	6209	880	16,820	281.9	408.7	585
106	41,177	348,992	2887	18,720	18,130	6149	897	16,940	280.1	409.3	613
114	33,267	310,555	2352	18,390	16,620	6139	764	16,870	273.2	393.6	605
122	33,889	303,373	2075	17,650	16,780	6093	818	17,270	280.5	387.5	612
134.5	32,182	291,935	1853	17,050	15,800	5776	828	16,840	264.6	388.0	593
139	38,420	325,717	899	17,280	14,470	6587	537	19,240	274.8	394.9	585
143	33,030	293,930	2734	17,980	32,740	5536	677	16,250	322.6	379.2	578
147	39,640	325,318	854	17,370	15,300	6315	590	19,560	268.1	397.9	580

(continued)

Table 2 (continued)

Depth (cm)	Al	Si	P	K	Ca	Ti	Mn	Fe	Sr	Zr	Ba
151	32,273	302,841	2396	17,530	25,880	5951	753	16,270	304.8	408.9	601
159	31,730	296,856	1372	17,520	24,600	5920	610	16,400	297.7	397.7	567
162.5	31,538	302,442	1493	17,190	25,510	5814	641	15,830	309.3	404.0	542
202	35,200	324,653	1973	17,350	20,000	6388	700	16,740	290.1	439.7	607
206	38,477	339,549	1702	17,700	27,730	6073	629	15,720	312.4	409.4	598
208	38,364	339,948	1662	17,430	27,690	6179	634	16,070	311.2	433.8	575
210	38,251	337,820	1615	17,720	26,720	5888	629	15,760	308.5	395.7	557
212	37,957	336,756	1661	17,480	27,740	5987	628	15,910	309.8	409.4	581
214	36,680	326,648	1666	16,970	28,190	5669	705	15,750	310.9	418.7	574
222	32,940	300,580	1229	16,790	20,930	6038	589	17,400	285.7	384.2	562
231	36,420	323,589	867	16,930	16,270	6167	589	18,490	262.6	378.8	538
255	35,324	309,757	948	16,690	16,120	6397	547	18,550	271.6	422.1	599
267	37,380	318,668	688	15,950	13,930	6510	536	18,730	252.9	405.8	591
284.5	36,985	302,708	477	15,650	14,260	6363	447	19,940	246.1	405.6	613
289	39,155	333,032	1872	17,400	22,160	5766	645	16,560	286.5	400.5	557
291	44,590	341,145	1206	17,830	17,400	6240	534	18,580	271.5	395.0	573
293	46,115	347,795	896	17,620	14,840	6237	491	18,880	259.6	444.8	591
295	46,025	341,278	813	17,330	13,550	6431	442	18,950	242.9	431.5	601
297	45,381	350,987	943	18,400	13,120	6256	485	18,480	255.0	432.1	668
308	35,007	313,747	1089	15,350	14,740	6083	566	17,660	265.3	387.2	619
317	34,872	313,215	721	15,340	10,880	6251	506	18,310	243.3	352.8	648
325	32,928	288,344	703	14,250	10,420	6342	669	18,650	252.8	416.5	634

(continued)

Table 2 (continued)

Depth (cm)	Al	Si	P	K	Ca	Ti	Mn	Fe	Sr	Zr	Ba
355	35,279	319,599	662	15,260	11,240	6151	676	17,430	253.8	400.5	574
363	34,906	319,599	460	14,560	9873	6324	552	17,480	249.1	418.3	638
371	36,827	318,136	935	15,950	13,230	6502	570	18,290	262.4	397.8	600
376	46,217	359,898	467	16,310	10,310	6358	510	17,100	239.2	408.2	640
377	45,991	356,972	422	16,630	10,840	6138	497	16,670	253.8	385.9	681
379	41,324	345,401	418	16,230	10,520	6090	486	16,650	238.9	361.3	564
381	45,392	357,637	412	16,410	9665	6045	519	16,130	233.5	402.8	698
383	45,573	364,287	431	16,470	10,360	6125	532	16,840	235.0	383.8	598
387	35,301	329,042	421	15,150	10,460	6151	531	16,550	245.3	400.8	620
395	38,669	322,525	216	14,670	9265	6463	477	18,460	229.1	391.4	637
399	39,923	298,053	220	14,350	10,270	6419	2450	20,550	241.8	413.8	680
	Al	Si	P_{av}	K	Ca	Ti	Mn	Fe	Sr	Zr	Ba
Average	**38209.11**	**325401.74**	**1444.24**	**16821.67**	**16282.83**	**6149.50**	**713.26**	**17372.78**	**266.33**	**405.70**	**609.37**
σ	**4247.42**	**20655.06**	**799.77**	**1150.01**	**5869.21**	**217.02**	**302.64**	**1169.49**	**25.16**	**19.63**	**37.01**
GRFQ02	s/d	s/d	18.3	13,690	9164	6274	587	15,390	235.7	456.8	523
GRFQ04	s/d	s/d	26.2	13,090	9408	5968	474	17,100	216.2	392.2	547
GRFQ05	s/d	s/d	3.6	13,360	9464	6359	634	15,800	239.2	464.4	635
GRFQ09	s/d	s/d	10.4	12,260	10,550	5902	915	20,140	213.9	376.6	533
GRFQ09 anthill	s/d	s/d	22.9	12,940	12,620	5665	457	18,990	222.8	369.5	592
GRFQ10	s/d	s/d	11.5	13,780	9956	6671	644	14,350	254.1	428.6	557
Average	–	–	**18.30**	**13186.7**	**10193.7**	**6139.8**	**618.5**	**16961.7**	**230.3**	**414.7**	**564.5**
σ	–	–	**7.47**	**559.3**	**1286.9**	**363.2**	**165.4**	**2230.5**	**15.5**	**41.1**	**42.0**

Table 3 Values of OM, carbonates, pH, and P_{av} from the mounds 5, 7, and 10, García Ricci site

Sample	Depth (cm)	% OM	% $C03^{=}$	pH	P_{av}	% P_{av}
Mound 5						
GR 05-1	56.5	5.23	s/d	5.7	1013	s/d
GR 05-2	78.5	5.04	s/d	6.3	489	s/d
GR 05-3	98.5	4.72	s/d	6.3	433	s/d
GR 05-4	118.5	4.58	s/d	6.5	740	s/d
GR 05-5	118.5	4.20	s/d	6.7	740	s/d
GR 05-6	145.5	4.05	s/d	6.7	1059	s/d
Mound 7						
GR7-1	32.0	4.69	1.20	5.48	1023	42.6
GR7-2	62.0	4.02	1.39	5.4	657	37.6
GR7-3	106.0	3.83	1.29	5.96	1389	52.1
GR7-4	122.0	4.62	1.33	6.27	966	41.1
GR7-5	151.0	4.76	1.62	6.88	1223	51.0
GR7-6	206.0	3.36	1.80	7.7	639	38.5
GR7-7	247.0	3.20	1.50	7.63	313	36.6
GR7-8	295.0	1.74	1.52	7.66	411	35.9
GR7-9	317.0	3.11	1.37	7.46	205	28.4
GR7-9[a]	355.0	s/d	s/d	7.56	509	76.9
GR7-10	377.0	1.95	1.17	7.58	176	40.8
Mound 10						
GR10-1	10.0	7.10	1.56	5.46	2477	s/d
GR10-2	36.5	5.20	1.54	5.98	2369	s/d
GR10-3	88.5	4.14	1.70	6.84	3777	s/d
GR10-4	133.5	4.94	3.28	7.33	1188	s/d
GR10-5	171.5	3.93	2.40	7.86	246	s/d
GR10-6	200.5	3.36	2.21	8.28	197	s/d
GR10-7	236.5	4.12	2.46	8.37	219	s/d
GR10-8	273.5	3.57	1.82	8.79	266	s/d
GR10-9	311.5	3.32	1.45	8.7	236	s/d
GR10-10	350.5	2.15	1.67	8.79	200	s/d
GR10-11	382.5	3.34	1.84	9.2	144	s/d
GR10-12	417.5	2.20	1.90	9.2	160	s/d

Table 4 Luminescence chronologies of the García Ricci site (Bracco et al. 2017c; Duarte et al. 2017)

Depth (m)	Fraction	Age BP	σ	Average weighted	σ weighted	Year AD/BC
Mound 05						
0.98	Silt	1644	136	–	–	370 AD
1.18	Silt	1800	127	–	–	220 AD
1.45	Silt	2188	420	–	–	170 BC
Mound 07						
0.72	Silt	2964	270	2910	205	894 BC
0.72	Silt	2830	315			
1.74	Silt	3448	320	3470	230	1454 BC
1.74	Silt	3494	330			
2.63	Silt	4633	390	4865	280	2849 BC
2.63	Silt	5111	400			
3.4	Silt	4865	395	5045	295	3029 BC
3.4	Silt	5267	440			
Mound 10						
0.78	Silt	2781	332	2610	270	593 BC
0.78	Silt	2543	270			
1.62	Silt	3412	335	3300	110	1283 BC
1.62	Silt	3222	122			
1.62	Sand	3314	326			
2.46	Silt	3668	381	3726	145	1709 BC
2.46	Silt	3890	318			
2.46	Sand	3598	164			
3.39	Silt	3748	373	3938	275	1922 BC
3.39	Silt	4023	410			
3.78	Silt	3614	239	3838	190	1822 BC
3.78	Sand	4071	328			
3.78	Silt	3819	343			
4.28	Silt	4135	400	3659	120	1643 BC
4.28	Sand	3676	228			
4.28	Sand	3562	141			

Table 5 ^{14}C chronologies of García Ricci site

Depth (m)		Age ^{14}C	Age ^{14}C cal BP (2 sigma) (Hogg et al. 2013)
Mound GR 05			
1.20	URU 0569	1640 ± 50 yr BP[a]	361–582 AD (p = 1)
1.50	URU 0575	2060 ± 50 yr BP[a]	194 BC–184 BC (0.009)
			150 BC–81 AD (0.978)
			98–111 AD (p = 0.013)
1.78	URU 0573	2620 ± 60 yr BP[a]	913–734 BC (0.810)
			689–662 BC (0.044)
			648–546 BC (0.146)
Mound GR 10			
4.14	Beta458464	3.680 ± 30 yr BP[b]	2.062–1.901 BC (0.822)
			2.134–2.079 BC (0.178)

[a]Bracco et al. (2017b)
[b]Duarte et al. (2017)

Results

Absolute Ages

Table 3 shows the new OSL and TL ages along with those available from the mounds 5, 7, and 10. Table 4 shows the ^{14}C dates available for mounds 5 and 10. The dates of mound 5 indicated that it was built during the period 900 BC–580 AD. The dates of mound 7 showed that it was built during the period 3000 BC–900 BC, and mound 10 during the period 1463 BC–590 BC. The set of dates indicate that the site was occupied continuously or recurrently for a period of at least 4500 years.

Elementary and Edaphic Analysis

Table 2 shows the results of the elementary analysis (XRF) of the matrix of mound 7. Table 5 presents the values for total phosphorus, available phosphorus, organic matter, carbonates, and pH in mounds 5, 7, and 10. The plotted data is presented in Fig. 6.

Research was focused on the elements of the matrix determined by XRF that presented a greater horizontal variability or that their concentrations were different from those of the samples of the plain. The elementary analysis of the matrix showed, as expected, Si and Al as dominant elements in the entire profile (Table 2). The concentration of both elements showed a high positive correlation ($R^2 = 0.742$). In the depths 35, 100, 200, 300, and 380 cm, an increase in the concentration of both is observed. The next most abundant elements are Fe, K, and Ca. Fe exhibits a marked vertical variability, with a greater concentration towards the base. K and Ca decrease towards the base, with a certain degree of correspondence between several of their peaks. Ba also shows higher concentrations in the mound matrix than in samples from the adjacent plain. Only two samples from the plain have Ba levels similar to those of the mound: GRFQ05 and GRFQ09. The first comes from a place that is very low altimetrically and, therefore, more frequently flooded. The second comes from an anthill.

Throughout the profile of mound 07, the total and available phosphorus (P_{tot} and P_{av}) showed a marked decrease (~2500 to ~200 ppm/~1500 to ~150 ppm) from the top to the base. A very high correlation between P_{tot} and P_{av} was also observed, in a ratio of approximately 2: 1 (see Fig. 6). The concentration of P_{av} for the three mounds varied between 4000 and 100 ppm, which even at the lower end is extremely high compared to the natural soils of the area: ~4 ppm (Hernández and Berger 2003: 36). The highest values of P_{av} appear approximately 200 cm deep in mounds 7 and 10. Mound 5 behaves similar to the first 170 cm of mound 7. In the samples of the plain surrounding the mounds P_{av} levels are notoriously lower, although higher than expected for natural soils (Hernández and Berger 2003: 36) with the exception of sample GRFQ05 (Table 2).

OM percentages of mounds 5, 7, and 10 were in the range of 4.05–5.23%, 1.8–4.7%, and 2.2–7.1%, with averages of 4.6%, 3.7%, and 3.9%, respectively. At mounds 7 and 10, the highest percentage of OM appeared at the highest levels, decreasing towards the base as the pH increases. In both mounds, when the pH reaches values close to 8, the OM percentage decreases to 3% or less. Mound 10 presented a correlation of R2 = 0.795 between both parameters. In mound 5, OM decreased towards the base and the pH increased, but both in a smaller proportion.

Table 6 presents the results of the P_{av} for the samples from natural soil and anthills, heated in the laboratory, using a muffle. For all samples, an increase in P_{av} is observed as the temperature increased, reaching concentrations of 3–10 times higher than the initial ones, with two maximums at 300 and 500 °C. The two soil samples from the plain showed a greater increase (from 13 and 8 ppm to 100 and 112 ppm).

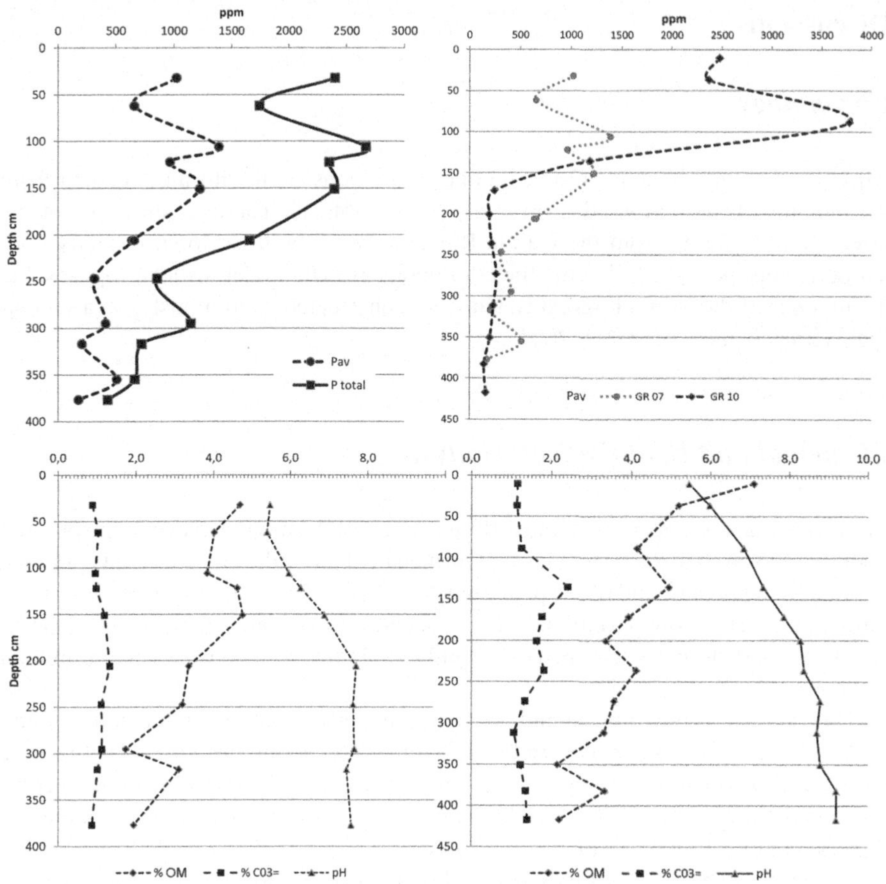

Fig. 6 Mound 7. vertical variation of P_{tot} and P_{av}. Mounds 7 and 10: vertical variation of P_{av}. Mound 7: vertical variation organic matter, carbonates, and pH. Mound 10: vertical variation organic matter, carbonates, and pH

Table 6 P_{av} for the samples from natural soil and anthills, heated with muffle

Temperature	–	100 °C	200 °C	300 °C	400 °C	500 °C	600 °C
Anthill	22	19	25	66	44	59	76
Anthill	25	20	22	68	44	57	74
Soil 0–15 cm	13	54	95	100	68	69	35
Soil 15–30 cm	8	30	32	49	46	112	70

Discussion

Chronology

All OSL, TL and ^{14}C dates (Tables 3 and 4) are consistent with the ^{14}C data from the mounds of the region (Bracco et al. 2015: Table 1). The data showed that the three mounds dating from the García Ricci site were built for long, partially contemporaneous periods, following the continuous growth model, which proposes that the mounds of the Merín Lagoon Basin were constructed continuously, on a secular scale (Bracco and Ures 1999, 2001).

Elemental and Edafological Analysis

The elementary composition of the different levels of the mound matrix is the product of the composition of the sediments that formed it, of the anthropogenic contributions, of other living beings, and of the pedogenic processes. This makes its interpretation a difficult task. However, variations in the concentration of some elements, due to their chemical characteristics, can be consistently explained as a consequence of human behavior.

The variations shown by Si and Al would indicate variations in the composition of the sediments that were used in the different construction instances. The positive correlation between both elements is consistent with a variation of the mineral composition of the sediments. This would indicate the use of sediments of heterogeneous mineral composition or of different origin.

The great vertical variability presented by Fe, Ca, and K does not appear as consistent with the mobilization processes of their compounds. It can also be explained by the variations in the composition of the sediments that were used in the different moments of construction of the mound. But the possibility that such variation is the consequence of variations in the quantity or composition of discarded materials during the occupation periods cannot be ruled out. For the first option, a certain correlation would be expected with the variations of Al and Si. However, said correlation is not observed.

If the levels of K and Ca of the matrix of the mound 7 are compared with those of the sediments of the plain adjacent to the mounds, it is observed that the first ones are significantly higher than the last ones (see the averages and errors in Table 2). These elements form the compounds, together with Si, Al, and P, more abundant in biomass ash (Krajnc 2015).

It has been observed that Ba concentrates in flooded soil with reducing conditions (Lima et al. 2012). Therefore, it would not be expected higher levels of Ba in the mound than in the sectors of the site most exposed to flooding. However, only two samples taken in the plains have Ba levels similar to those of the mound, the rest have lower levels. Those that show similar levels come from a sector that flood with

Fig. 7 Concretions of burned earth. The arrows indicate holes and depressions whose characteristics coincide with galleries of anthills of *Camponotus punctulatus*

greater frequency and from an anthill of *Camponotus punctulatus*. These ants build their epigeal nests with material that they transport from the lowest levels of the soil (Cosarinsky 2003). Therefore, the levels of Ba would indicate the use of sediments from flooded soil with reducing conditions and/or anthills in the construction of the mounds. Bracco et al. (2000c, 2017b) have indicated that fragments of anthills were used to make the concretions of the burned earth. As noted, these concretions can be up to 30% of the mound matrix (Fig. 7).

Two of the main parameters that affect the processes that cause phosphates to migrate in the profile of a soil are pH and organic matter content (Holliday and Gartner 2007). It is observed in the mounds 7 and 10 that as the P_{av} levels decreased, the organic matter (OM) decreased and the pH increased from neutral to basic (see Fig. 6, Table 5). On mound 5, the same trend was observed, but the OM % did not become as low, nor the pH so high. Its variations of P_{av}, OM % and pH are similar to those of the upper 170 cm of mounds 7 and 10, in the zone of greatest influence of vegetation. This leads to think that the vertical distribution of P_{av}, that it is observed today would be the consequence of pedogenic processes that modified an anthropogenic record (Holliday and Gartner 2007: 305). The P_{av} values shown by the samples taken in the plain—with one exception—are much lower than the P_{av} values of the mounds, but they are above the natural threshold. This may be due to contemporary human activity to the construction/occupation of the mounds and/or to the transport of the mound matrix to the surrounding areas by animals.

The OM % of the matrix of the mounds of the García Ricci site are higher than those observed at site CH2D01 (average = 2.5%, Table 1) and similar to those of the superficial horizons of the non-flooded soils of the region, which vary from 3.5 to 5.7% (Durán 1990: 21). Likewise, the highest values of mounds are similar or lower than those reported for the superficial horizons of the wetlands of India Muerta: 9.7% and 5.5% (Durán 1990). On the other hand, the pH of the mounds 07 and

10 presented similar values in its upper part, around 5.5. They increase to the base reaching values of 7.6 and 9.2, respectively. Similar trends are observed in the profiles of cultivated and natural soils of the region (Durán 1990: Fig. 16). The OM/pH ratio showed another similarity with the regional natural soils (see Durán 1990). The correlation between both parameters is consistent with the pedogenic processes, which would have accelerated once the mounds have stopped being built and/or have been colonized by the vegetation. If the interpretation is correct, it must be admitted that the original amount of OM was lower than what it may be observed today. They would have been similar to those present at the base of the highest mounds. In any case, the OM levels confront us with an apparent inconsistency. Durán (1990: 24–25) observed that, for the site CH2D01, the level of P_{av} is much higher than that of natural soils throughout the country and even higher than that of heavily fertilized soils. This author added that everything indicated that the remarkably high levels of available phosphorus, as well as high levels of K, are the result of anthropogenic contributions of large amounts of organic waste over a sustained period of time. The levels of P_{av} at the Garcia Ricci site are even higher than that of the CH2D01 site. But the contribution of large amounts of organic waste was not reflected in the OM %. As it has been mentioned, the amount of OM is similar (or less) than that of natural soils (see above). The low percentages of OM can be explained if it is admitted that OM has been degraded by fire, leaving only ash and coal. The ashes and the heating of the sediments may be the origin, although not necessarily in its entirety, of the high levels of P_{av}. The use of fire not only solves the previous inconsistency but also explains other properties of the mound matrix.

In the first place, the presence of burned sediment ("tierra quemada") proves the use of fire. Its abundance allows to estimate how intense was the use of fire, the amount of vegetable fuel used and waste produced. The luminescent spectra of the quartz fraction, observed during TL dating of the burned sediment, indicate that its temperature reached at least 360 °C. Starting from the specific heat of the dry soil (0.44 kcal/kg °C), it may be estimated that 158.4 kcal are required for one sediment kg to reach that temperature. If the average calorific value of hard and soft wood is taken into consideration (~2000 kcal/kg; Krajnc 2015) it may be inferred that it was necessary to burn 0.079 kg of vegetable fuel per kg of sediment. But the result is underestimated, since it has not been considered, among other things, that the sediment has water and that not all the heat produced is transferred to the sediment. For experimental tests of soil thermal alteration, Olarieta et al. (2011) have reported the use of 0.120 kg of plant fuel per kg of soil (vide infra). If the average value (theoretical and experimental) for fuel requirements (0.0995 kg per kg of soil) is taken into account, it may be estimated that the process of raising a mound of 4 m in height (~6000 m^3), with 30% sediment burned in its matrix (~1800 m^3)—similar to mounds 7 and 10 of the García Ricci site—demanded approximately 411,000 kg of plant fuel. But again, the calculation is underestimated, because not only the concretions of the burnt sediments showed evidence of exposure to heat. The TL ages obtained from the fine fraction of the mound matrix are not statistically differentiable from the TL ages obtained from the burned earth (Table 3) (Bracco et al. 2017a, c). This showed that all (or almost all) of the matrix reached high temperatures.

Fire has been used to modify the properties of soils, mainly by traditional agricultural societies. The fire can cause the soil to increase the availability of nutrients (mainly P, K, and Ca), increase the pH, lose its structure, and modify its texture and its color (Santín and Doerr 2016). A singular modality of these practices was used in the Iberian Peninsula until the adoption of industrial fertilizers in the mid-twentieth century. It consists on the construction of "*formiguers*" (anthills in Catalan, name given by its morphological similarity). *Formiguers* are structures similar to brick oven, which were mainly used to increase soil fertility from burning biomass. Their use was widely extended and imposed during the seventeenth and eighteenth centuries for some regions of Catalonia (Mestre Artigas and Mestres Jané 1949). Vegetable material with low heat output but easy to obtain was used as fuel: crop residues, pruning branches, and weeds (Mestre Artigas and Mestres Jané 1949). The plant material is stacked and then surrounded and covered with clods. Finally, the spaces between the clods are filled with crushed earth to reduce air intake and heat dissipation. They reached heights of 0.8 m and diameters of 1.5 m in Mallorca (Vadel et al. 2010); its volume varied between 0.5 and 2.0 m^3 (Olarieta et al. 2011). The *formiguers* burned for 4 or 5 days. The soil acquired more vivid colors where it reached higher temperatures (Mestre Artigas and Mestres Jané 1949; Vadel et al. 2010). Experimental tests have shown that these ovens can reach temperatures above 300 °C in the core, using 68 kg of plantfuel and 550 kg of soil (Olarieta et al. 2011). The *formiguers* were dismantled and scattered once the fire is extinguished (Fernández-Cano and Togores 2011: 1301).

Experimentation showed that the practice decreases the organic carbon levels of the soil, leaves traces of carbon, and increases the exchangeable K and the labile and total P. Similar practices of soil thermal alteration are used nowadays in India and Bhutan. In India (Meghalaya), the technique shows some variations. The fuel (bushes, grass, and low tree branches) is arranged in lines, and once it is dry, it is covered with soil, and then set on fire (Olarieta et al. 2011). The soil thermal alteration practices were not only appreciated for the increase of nutrients but also as a means to disinfect the soils cover, eliminate unwanted insects and weeds, as well as to reduce the biological acidity and modify the permeability and texture of soils (Olarieta et al. 2011; Santín and Doerr 2016). In the latter case, this is due to the production of silt and sand-sized particles from the fusion of clay particles (Fernández-Cano and Togores 2011; Olarieta et al. 2011). The ashes also contribute to increasing the slime fraction.

Not only the geochemical characteristics and the present of burned earth of the mound matrix can be explained through the use of soil thermal alteration practices similar to those that have been described. These practices also explained: (1) the presence of combustion evidence and the almost absence of combustion structures in the mounds (Bracco 2006); (2) the presence in the matrix of scattered coal remains that generally correspond to pieces of small branches and palm nuts (Inda 2003); (3) a "lighter" texture of the matrix, with smaller amounts of clay than the soil of the region (Fig. 8); and (4) the way of production of burned earth.

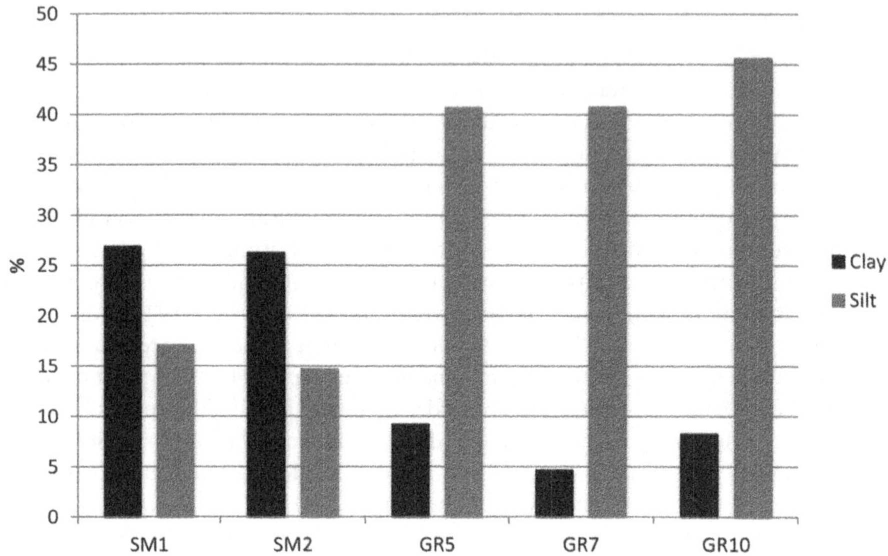

Fig. 8 Percentages of silt and clay for two natural soils of medium and low plains (SM1 and SM2) (Durán 1990) and mounds 05, 07, and 10 of the García Ricci site

Conclusions

The investigations on the processes of formation of the mounds supposed that it was mainly the product of the accumulation of sediments and debris. The data provided by the elemental analyses provided information on the possible sources of the sediments used for the construction of the mounds. They also indicate that possibly the construction process included organic matter. The data provided by edaphic analyses showed that the anthropogenic record was possibly altered by pedogenic processes. But the conjugation of the analytical data with other characteristics of the mound matrix has led the present authors to propose that the processes of construction of the mounds not only included accumulation of sediment and debris.

The high levels of phosphates in the matrix of the mounds indicated a contribution of large amounts of organic matter. On the other hand, the relatively low levels of organic matter in the matrix of the mounds indicated that it would have been degraded by natural and artificial processes. Different indications allowed to consider as probable that the degradation of OM was the consequence of artificial processes that integrated the use of fire.

The abundance of burned earth demonstrated the extensive use of fire. A theoretical calculation of the amount of heat needed to burn earth gave a notion of the amount of plant fuel that would have been needed to build a 4-m-high mound, in which the burned earth is 30% of its matrix: 411,000 kg. This large burning would necessarily produce considerable amounts of charcoal and ash, the latter quite rich, among other elements, in P and K.

This research refers to traditional practices to illustrate how the process of thermal alteration of the sediments could have been, and the consequences and motives for their use. The use of these practices not only explained the geochemical characteristics of the matrix and the presence of the burned earth in it, but also account for: (1) the presence of evidence of combustion and the near-absence of combustion structures in the mounds, (2) the presence in the matrix of scattered charcoal remains that generally correspond to small pieces, (3) a "lighter" texture of the matrix, with smaller amounts of clay than the soil of the region, and (4) the mode of production of the burned earth.

Thermal alteration practices modify the soil by increasing the availability of some nutrients, disinfecting and cleaning the fields, eliminating unwanted living beings (plants and animals) as well as changing the structure, texture and permeability of soils. The repetition of these practices in one place will produce a higher and very different substrate from the natural substrate of the alluvial plains. A higher place with a permeable substrate in the middle of a flood plain is a very favorable place for humans and nonhumans to carry out multiple activities. In the structure of a landscape, these places become patches. Paying attention to the long periods of formation of the mounds and mainly emphasizing that they are necessarily the consequence of a recursive behavior, it has been proposed to designate them as "recursive patches". The two main characteristics of a recursive patch are, on the one hand, being part of the landscape (the product of human-environment interaction) and, on the other, its character of emergent phenomena. Usually, the consequences of recursive behaviors are of a different order from the products that were searched each time each behavior that integrates them was executed. Many social phenomena (and their material consequences) are not the result of a conscious project with defined objectives but are generated by traditions. Erickson (2006: 125) sustained that people create, transform, and manage the environments in which they live and prosper through their culture, accumulating knowledge and multigenerational management practices (the emphasis is of the present authors). The population and the chronologies of the mounds of the Paso Barrancas–India Muerta area indicated the temporospatial scale of the landscape transformation that was produced by the implantation of those recursive patches. Evidence has been provided that this modification of the landscape was not achieved only by accumulating sediments and waste but also would have been using pyrotechnology.

Acknowledgements This work was done within the framework of the project "Use of Nuclear Techniques in Support for the Conservation and Preservation of Objects of Cultural Heritage" (RLA/0/058—2016-17), IAEA-UN. We thank Daniel Loponte for the critical reading of the first draft and Laura del Puerto and Joaquín Mazzarino for their collaboration in field tasks.

References

Baeza J, Panario D (1999) La horticultura indígena en las estructuras monticulares. In: Actas de las Primeras Jornadas sobre Cenozoico en Uruguay, 20–21 Dec 1999. SUG, INGEPA and UNCIEP, Facultad de Ciencias, Montevideo, Uruguay

Balée WL, Erickson CL (eds) (2006) Time and complexity in historical ecology: studies in the neotropical lowlands. Columbia University Press, New York, U.S.A.

Bracco R (2006) Montículos de la cuenca de la laguna Merín: Tiempo, espacio y sociedad. Latin Am Antiq 17:511–540. http://sci-hub.tw/10.2307/25063070

Bracco R, Ures C (1999) Ritmos y dinámica constructiva de las estructuras monticulares. Sector sur de la cuenca de la laguna Merín. Uruguay. In: López JM, Sanz M (eds) Arqueología y Bioantropología de las Tierras Bajas. Facultad de Humanidades y Ciencias de la Educación, Universidad de la República, Montevideo, Uruguay, pp 13–33

Bracco R, Cabrera L, López JM (2000a) La prehistoria de las tierras bajas de la cuenca de la laguna Merín. In: Duran A, Bracco Boksar R (eds) Arqueología de las Tierras Bajas. Ministerio de Educación y Cultura, Comisión Nacional de Arqueología, Montevideo, Uruguay, pp 13–38

Bracco R, Fregeiro MI, Panarello H, Odino R, Souto B (2000b) Dieta, modos de producción de alimentos y complejidad. In: Duran A, Bracco Boksar R (eds) Arqueología de las Tierras Bajas. Ministerio de Educación y Cultura, Comisión Nacional de Arqueología, Montevideo, Uruguay, pp 227–248

Bracco R, Montaña JR, Nadal O, Gancio F (2000c) Técnicas de construcción y estructuras monticulares, termiteros y cerritos: de lo analógico a lo estructural. In: Duran A, Bracco Boksar R (eds) Arqueología de las Tierras Bajas. Ministerio de Educación y Cultura, Comisión Nacional de Arqueología, Montevideo, Uruguay, pp 285–300

Bracco R, Ures C (2001) Ritmo y dinámica constructiva de las estructuras monticulares. sector sur de la cuenca de la Laguna Merín. In: Arqueología Uruguaya Hacia el Fin del Milenio. Anales del IX Congreso Nacional de Arqueología, 16–19 June 1997. Asociación Uruguaya de Arqueología, Colonia del Sacramento, Uruguay, pp 41–53

Bracco R, del Puerto L, Inda H, Castiñeira C (2005) Mid-late Holocene cultural and environmental dynamics in Eastern Uruguay. Quatern Int 132:37–45. http://sci-hub.tw/10.1016/J.QUAINT.2004.07.017

Bracco R, del Puerto L, Inda H (2008) Prehistoria y arqueología de la cuenca de Laguna Merín. In: Loponte D, Acosta A (eds) Entre la Tierra y el Agua. Arqueología de Humedales de Sudamérica, AINA, Buenos Aires, pp 1–59

Bracco R, del Puerto L, Inda H, Panario D, Castiñeira C, García-Rodríguez F (2011) The relationship between emergence of mound builders in SE Uruguay and climate change inferred from opal phytolith records. Quatern Int 245:62–73. http://sci-hub.tw/10.1016/j.quaint.2010.05.025

Bracco R, Inda H, del Puerto L (2015) Complejidad en montículos de la cuenca de la laguna Merín y análisis de redes sociales. Intersecciones en Antropología 16:271–286

Bracco R, Duarte C, Bazzino A (2017a) Estructuras monticulares: modificación del paisaje a través de la creación de parches por medio de la termoalteración de sedimento. In: Jornadas Académicas de FHCE «Profesor Washington Benavídez» (VII Jornadas de Investigación; VI Jornadas de Extensión y V Encuentro de Egresados y Estudiantes de Posgrado), 11–13 Oct 2017, GT 31: Investigación arqueológica en Uruguay. Montevideo, Uruguay

Bracco R, Duarte C, Bazzino A, Panario D (2017b) Estructuras monticulares y pirotecnología. Séptimo Encuentro de Discusión Arqueológica del Nordeste Argentino, 7 EDAN. Centro de Investigaciones Científicas y Transferencia de Tecnología a la Producción - CONICET, Diamante. Entre Ríos, Argentina

Bracco R, Duarte C, Bazzino A, Panario D (2017c) Estructuras monticulares, hormigueros y formiguers. La seducción de la analogía. In: Séptimo Encuentro de Discusión Arqueológica del Nordeste Argentino, 7 EDAN. Centro de Investigaciones Científicas y Transferencia de Tecnología a la Producción - CONICET, Diamante, Entre Ríos, Argentina

Bray RH, Kurtz LT (1945) Determination of total, organic and available forms of phosphorus in soils. Soil Sci 59:39–46. http://sci-hub.tw/10.1097/00010694-194501000-00006

Browne Ribeiro AT (2011) Acts, practices, and the creation of place: Geoarchaeology of a Terra Preta de Índio site in the Central Amazon. Unpublished doctoral dissertation (PhD), University of California

Cabrera L (1999) Funebria y sociedad entre los 'Constructores de Cerritos' del Este uruguayo. In: López JM, Sanz M (eds) Arqueología y Bioantropología de las Tierras Bajas. Facultad de Humanidades y Ciencias de la Educación, Universidad de la República, pp 63–80

Cabrera L (2013) Construcciones en tierra y estructura social en el Sur del Brasil y Este de Uruguay (ca. 4.000 a 300 a. A.P.). Techne 1:25–33

Castiñeira C, Piñeiro G (2000) Análisis estadístico textural para el estudio de las columnas estratigráficas de la excavación I y II del Bañado de los Indios. In: Duran A, Bracco Boksar R (eds) Arqueología de las Tierras Bajas. Ministerio de Educación y Cultura, Comisión Nacional de Arqueología, Montevideo, Uruguay, pp 467–478

Castiñeira C, Capdepont I, del Puerto L, Blasi AM (2015) Aportes de la geoarqueología para el estudio de la variabilidad constructiva prehispánica de cerritos del este uruguayo y delta del rio Paraná. In: Rubin De Rubin JC, Favier Dubois CM, da Silva RT (eds) Geoarqueologia na América do Sul. Editora da PUC Goiás, Goiânia, Brasil, pp 55–90

Copé SM (1991) A ocupação pré-Colonial do Sul e Sudeste do Rio Grande do Sul. In: Jacobus AL, Kern AA (eds) Arqueologia pré-histórica do Rio Grande do Sul. Mercado Aberto, Porto Alegre, Rio Grande do Sul, Brasil, pp 191–219

Cosarinsky MI (2003) Micromorfología de nidos de termitas de la República Argentina. Unpublished doctoral thesis, Facultad de Ciencias Exactas y Naturales, Universidad de Buenos Aires

Curbelo C, Cabrera L, Fusco N, Martínez E, Bracco R, Femenías J, López JM (1990) Sitio CH2D01, área de San Miguel, Depto. de Rocha, RO del Uruguay. Estructura de sitio y zonas de actividad. Revista do CEPA 17:333–344

del Puerto L (2011) Ponderación de recursos vegetales silvestres del este del Uruguay: rescatando el conocimiento indígena tradicional. Trama Revista de Cultura y Patrimonio 3:22–41

del Puerto L (2015) Interrelaciones humano-ambientales durante el Holoceno tardío en el este del Uruguay: Cambio Climático y Dinámica Cultural. Unpublished doctoral thesis, PEDECIBA, Universidad de la República. Montevideo

Duarte C, Bracco R, Panario D, Tassano M, Cabrera M, Bazzino A, del Puerto L (2017) Datación de estructuras monticulares por OSL/TL. Revista de Antropología del Museo de Entre Ríos 3:14–26

Durán A (1990) Observaciones sobre los suelos del sitio arqueológico de San Miguel. Unpublished technical report. Montevideo, Uruguay

Erickson CL (2006) The domesticated landscapes of the Bolivian Amazon. In: Balée WL, Erickson CL (eds) Time and complexity in historical ecology: studies in the neotropical lowlands. Columbia University Press, New York, U.S.A., pp 235–278

Erickson CL (2008) Amazonia: the historical ecology of a domesticated landscape. In: Silverman H, Isbell W (eds) The handbook of South American Archaeology. Springer, New York, NY, U.S.A., pp 157–183

Fernández-Cano LH, Togores JH (2011) Tratado de viticultura, 4th edn. Mundi-Prensa Libros, Madrid, Spain

Forman RTT, Godron M (1986) Landscape ecology, 1st edn. Wiley

García-Rodríguez F, Stutz S, Inda H, del Puerto L, Bracco R, Panario D (2010) A multiproxy approach to inferring Holocene paleobotanical changes linked to sea-level variation, paleosalinity levels, and shallow lake alternative states in Negra Lagoon, SE Uruguay. Hydrobiologia 646:5–20. http://sci-hub.tw/10.1007/s10750-010-0184-0

Gianotti C (2000) Paisajes monumentales en la región meridional sudamericana. Gallaecia 19:43–72

Gianotti C, Criado Boado F, López JM (2008) Arqueología del paisaje: la construcción de cerritos en Uruguay. Informes y Trabajos (Dedicado a: Excavaciones en el exterior 2007) 1:177–185

Heiri O, Lotter AF, Lemcke G (2001) Loss on ignition as a method for estimating organic and carbonate content in sediments: reproducibility and comparability of results. J Paleolimnol 25:101–110. http://sci-hub.tw/10.1023/A:1008119611481

Hernández J, Berger A (2003) Dinámica del fósforo en sistemas de arroz - pasturas: caracterización de parámetros de suelos para estimar la disponibilidad de fósforo: indicadores de disponibilidad de fósforo y respuesta del cultivo de arroz a la fertilización fosfatada. In: Investigaciones agronómicas: Reporte Técnico Anual 2003: área cultivos. Programa Nacional de Arroz. Instituto Nacional de Investigación Agropecuaria, Treinta y Tres, Uruguay, pp 34–40

Hogg AG, Hua Q, Blackwell PG, Niu M, Buck CE, Guilderson TP, Heaton TJ, Palmer JG, Reimer PJ, Reimer RW, Turney CSM, Zimmerman SRH (2013) SHCal13 southern hemisphere calibration, 0–50,000 Years cal BP. Radiocarbon 55:1889–1903. http://sci-hub.tw/10.2458/azu_js_rc.55.16783

Holliday VT, Gartner WG (2007) Methods of soil P analysis in archaeology. J Archaeol Sci 34:301–333. http://sci-hub.tw/10.1016/J.JAS.2006.05.004

Inda H (2003) Antracología: Paleoetnobotánica del fuego en la Prehistoria de la Región Este. Puntas de San Luis, Rocha, Uruguay. Tesis Taller II en Arqueología. Licenciatura en Ciencias Antropológicas. Facultad de Humanidades y Ciencias de la Educación. Universidad de la República, Montevideo

Inda H (2011) Paleolimnología de cuerpos de agua someros del Sudeste del Uruguay: Evolución Holocena e impacto humano. Editorial Académica Española, Spain

Iriarte J (2003) Mid-Holocene emergent complexity and landscape transformation: the social construction of early formative communities in Uruguay, La Plata basin. Unpublished doctoral thesis (PhD), University of Kentucky

Iriarte J, Holst I, López JM, Cabrera L (2000) Subtropical wetland adaptation in Uruguay during the mid-Holocene: an archaeobotanical perspective. In: Purdy BA (ed) Enduring records: the environmental and cultural heritage of wetlands. Oxbow Books, Oxford, pp 61–70

Krajnc N (2015) Wood fuels handbook. Food and Agricultural Organization of the United Nations, Pristina, Kosovo

Lima ESA, Amaral Sobrinho NMB do, Magalhães MOL, Guedes J do N, Zonta E (2012) Absorção de bário por plantas de arroz (Oryza sativa L.) e mobilidade em solo tratado com baritina sob diferentes condições de potencial redox. Química Nova 35:1746–1751. http://sci-hub.tw/10.1590/s0100-40422012000900008

López JM (2001) Las estructuras tumulares (cerritos) del litoral atlántico uruguayo. Latin Am Antiq 12:231–255. http://sci-hub.tw/10.2307/971631

López JM, Criado Boado F, Gianotti C (2001) Arqueología en el valle del arroyo Yaguarí. Diseño de proyecto y primeras actividades. In: Beovide L, Barreto I, Curbelo C (eds) Actas del X Congreso Nacional de Arqueología. La arqueología uruguaya entre los desafíos del nuevo siglo (CD-Rom). Asociación Uruguaya de Arqueología, Montevideo, Uruguay

Mestre Artigas C, Mestres Jané A (1949) Aportación al Estudio de la Fertilización del Suelo por Medio de Hormigueros. Boletín del Instituto Nacional de Investigaciones Agronómicas 109:125–163

Milheira RG, Garcia AM, Ribeiro BLR, Ulguim PF, Da Silveira CS, Sanhudo MDS (2016) Arqueologia dos Cerritos na Laguna dos Patos, Sul do Brasil: uma síntese da ocupação regional. Revista Cadernos do Ceom 29:33–63. http://sci-hub.tw/10.22562/2016.45.02

Mut P (2015) Paleodieta de los pobladores prehistóricos del este del Uruguay: un retrato isotópico. Anuario de Arqueología, Instituto de Ciencias Antropológicas Departamento de Arqueología – Facultad de Humanidades y Ciencias de la Educación – UdelaR, pp 147–178

Olarieta JR, Padrò R, Masip G, Rodríguez-Ochoa R, Tello E (2011) 'Formiguers', a historical system of soil fertilization (and biochar production?). Agr Ecosyst Environ 140:27–33. http://sci-hub.tw/10.1016/J.AGEE.2010.11.008

Panario D, Gutiérrez O, Achkar M, Bartesaghi L, Ceroni M (2015) Clasificación y mapeo de ambientes de Uruguay. In: Brazeiro A (ed) Eco-Regiones de Uruguay: Biodiversidad, Presiones y

Conservación. Aportes a la Estrategia Nacional de Biodiversidad, Facultad de Ciencias, CIEDUR, VS-Uruguay, SZU, Montevideo, Uruguay, pp 32–45
PROBIDES (1999) Plan Director. Reserva de Biosfera Bañados del Este/Uruguay, Unión Europea, PNUD, GEF, Rocha, Uruguay
Reimer PJ, Bard E, Bayliss A, Beck JW, Blackwell PG, Ramsey CB, Buck CE, Cheng H, Edwards RL, Friedrich M, Grootes PM, Guilderson TP, Haflidason H, Hajdas I, Hatté C, Heaton TJ, Hoffmann DL, Hogg AG, Hughen KA, Kaiser KF, Kromer B, Manning SW, Niu M, Reimer RW, Richards DA, Scott EM, Southon JR, Staff RA, Turney CSM, van der Plicht J (2013) IntCal13 and Marine13 radiocarbon age calibration curves 0–50,000 Years cal BP. Radiocarbon 55:1869–1887. http://sci-hub.tw/10.2458/azu_js_rc.55.16947
Santín C, Doerr SH (2016) Fire effects on soils: the human dimension. Philos Trans R Soc Lond Ser B Biol Sci 371:art 20150171. http://sci-hub.tw/10.1098/rstb.2015.0171
Santos FD, Lopes Mota A, Moniz G, Ramos L, Taborda R (2015) As recomendações do relatório do Grupo de Trabalho do Litoral – 2014 e a sua aplicação. In: VIII Congresso sobre Planeamento e Gestão das Zonas Costeiras dos Países de Expressão Portuguesa. Aveiro, Portugal, pp 1–9
Schmitz PI (1976) Sítios de Pesca Lacustre em Rio Grande, RS, Brasil. (Tesis obtención del grado de 'Livre-docente', por la Universidade Católica do Rio Grande do Sul, PUCRS), Instituto Anchietano de Pesquisas, Universidade do Vale de Rio dos Sinos, São Leopoldo, Brasil
Schmitz PI (1981) Contribuciones a la prehistoria de Brasil. Pesquisas, Série Antropologia 32:1–243
Terrell JE, Hart JP, Barut S, Cellinese N, Curet A, Denham T, Kusimba CM, Latinis K, Oka R, Palka J, Pohl MED, Pope KO, Williams PR, Haines H, Staller JE (2003) Domesticated landscapes: the subsistence ecology of plant and animal domestication. J Archaeol Method Theory 10:323–368. http://sci-hub.tw/10.1023/B:JARM.0000005510.54214.57
Vadel J, Marquès M, Amengual D, Roca A, Adrover M, Farrús E, Martorell A (2010) Evaluación agronómica de los hormigueros. In: Actas completas del IX Congreso SEAE "Calidad y seguridad alimentaria", 6–9 Oct 2010. Lleida, Spain, pp 1–7

Record of Late Holocene Human Occupations in Coastal Deposits of the Middle Uruguay River

Carola Castiñeira Latorre, Rodrigo Costa Angrizani, Eduardo Apolinaire, Marco Alvarez, Irina Capdepont, Adriana Blasi, Laura Maravilla, Florencia Mari and Michael Zech

Abstract The archaeological record recovered in fluvial deposits of the Uruguay River has been widely used to build interpretative frameworks about the South American prehistory. The interpretation of these records helped to consolidate the North American Clovis First paradigm and, currently, to contribute to its discussion. They are useful as well to understand the technological divergences of lithic, pottery, and agricultural pre-Hispanic industries on a continental scale. The information of this sector mostly corresponds to the fieldwork carried out by the Archaeological Rescue

C. Castiñeira Latorre (✉)
CONICET- Museo Argentino de Ciencias Naturales "Bernardino Rivadavia", Av Angel Gallardo, 470 C1405DJR Buenos Aires, Argentina
e-mail: carolacl2004@gmail.com

R. C. Angrizani
CONICET - Centro de Investigaciones Científicas y Transferencia de Tecnología a la Producción (UADER), Materi y España s/n, 3105 Diamante, Entre Ríos, Argentina

E. Apolinaire
CONICET - Facultad de Ciencias Naturales y Museo (UNLP), Paseo del Bosque s/n, 1900 La Plata, Argentina

M. Alvarez · L. Maravilla
FCNyM - División de Mineralogía, Petrología y Sedimentología - Facultad de Ciencias Naturales y Museo (UNLP), Paseo del Bosque s/n, 1900 La Plata, Argentina

I. Capdepont
Laboratorio de Estudios del Cuaternario (MEC-UNCIEP, IECA Facultad de Ciencias) - Centro Universitario Regional del Este (CURE-Rocha), Universidad de la República, Iguá 4225, 11400 Montevideo, Uruguay

A. Blasi
CIC - División de Mineralogía Petrología y Sedimentología - Facultad de Ciencias Naturales y Museo (UNLP), Paseo del Bosque s/n, 1900 La Plata, Argentina

F. Mari
LATYR - Laboratorio de Radiocarbono, Centro de Investigaciones Geológicas. CONICET-UNLP, 113 N°275, B1904DPK La Plata, Buenos Aires, Argentina

M. Zech
Landscape Ecology, Institute of Geography, Technical University of Dresden, Dresden, Germany

H. Inda Ferrero and F. García Rodríguez (eds.), *Advances in Coastal Geoarchaeology in Latin America*, The Latin American Studies Book Series,
https://doi.org/10.1007/978-3-030-17828-4_8

Mission of Salto Grande before the beginning of the construction of the hydroelectrical dam in the middle Uruguay River. Recently, an international and interdisciplinary team of archaeology begun to develop systematic investigations in order to recognize and actualize the information of this area. One of the main goals of this work is to reconstruct the paleoecological scenario in which the pre-Hispanic populations lived and interacted with each other and with the environment. To this purpose, different stratigraphic successions on the middle littoral Uruguay River were surveyed. They allowed to define and date a succession of fluvial erosive–aggradational episodes that characterized the fluvial dynamics during the Late Holocene. These episodes can be linked to the evolution of paleoclimatological phenomena supported by regional models. On the other hand, the archaeological records registered in the successions of the studied sites allowed to understand the coast of the Uruguay River as a hot spot area since its early human colonization.

Keywords Uruguay river · Holocene · Paleoenvironmental evolution · Pre-Hispanic settlement pattern

Introduction

The fluvial littoral—riparian and neighboring zones—has constituted substantial areas for the development of human populations. Lands in contact with waters or under its influence have high biodiversity and have been strategic areas for the development and maintenance of social connectivity (Arzamedia et al. 2015; Capdepont et al. 2017; González del Tánago and García de Jalón 1998). According to the archaeological literature addressing southern South America, the coast of the Uruguay River (Fig. 1) constitutes an example of the cultural value of these environments (e.g., Boretto et al. 1973; Bórmida 1964; Bueno and Dias 2015; Cabrera 1994a, b; Cerutti and Gonzáles 2007; Cigliano 1968, Diaz 1977; Guidón 1984, 1989; Hilbert 1985; Rodríguez 1992; Taddei 1964, 1980).

The coastal areas of the Uruguay River began to be populated during the Pleistocene—Holocene transition by groups of hunter-gatherers (e.g., Borrero et al. 1998; Hilbert 1991; Politis et al. 2004; Suárez et al. 2017) and since then, it has constituted a continuous scenario in the sociocultural development of the region (e.g., Cerutti 2013; Guidón 1984, 1989). The archaeological evidence, conserved in the coastal deposits of the Uruguay River, substantially contributed to the Clovis First paradigm and more recently to its discussion (see Politis et al. 2015, 2016).

The results obtained by the archaeological investigations developed in the coastal zone of the Uruguay River are added to those achieved by the archaeological activities that took place in the territories associated to the major rivers of the South American lowlands, such as the Amazon, Orinoco, Pilcomayo, Bermejo, Paraguay, and Paraná. These results have been crucial for the recognition and discussion of the multi-direction aspect of the human colonization process on a continental scale, as well as the chronology and dispersion of cultural innovations such as ceramic tech-

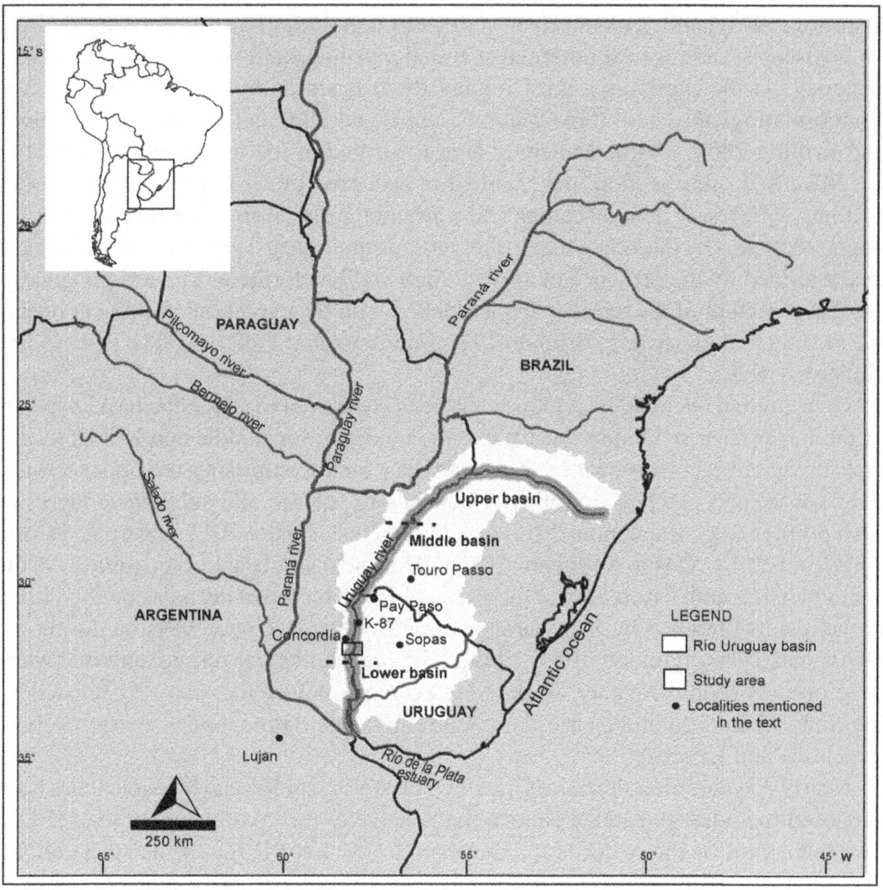

Fig. 1 Locations and sites mentioned in the text. Limits of the upper, middle, and lower basins

nology, agriculture, earth architecture, among other cultural knowledge (Berberian and Nielsen 2001; Gónzales-Ruibal et al. 2011; Morcote Ríos et al. 2006; Iriarte et al. 2017; Politis and Bonomo 2012; Suárez 2015). In this regard, the work intends to be a new contribution to this knowledge, by presenting the geoarchaeological results that have been recently achieved in the coastal zone of the middle basin of the Uruguay River (MBUR). First, it should be acknowledged the depositional dynamics that affect coastal fluvial deposits in our study area in order to contribute to the background knowledge. In this way, the beginnings of the alluvial geoarchaeology for the south of South America will be defined (Ferring 2017).

The headwaters of the Uruguay River are located at the confluence of the Pelotas and Canoas rivers, in the area of the Serra Geral National Park—Brazil (27°36′26″S–51°27′24″W). The river then flows toward the south, along 1800 km to its mouth in the Río de la Plata estuary (*circa* 34°14′14.32″S–58°2′18.64″W). The hydrological dynamics of the river varies throughout its basin generating particular landscape features in southernmost Brazil, north-eastern Argentina, and western Uruguay. At the same time, the Uruguay River basin is divided, geologically and geomorphologically, into three sections: upper, middle, and lower basins (Iriondo and Kröling 2004, 2008). The study area is located in the littoral western zone of the MBUR. In this section, the riverbed is occupied by several islands and rocky bottoms with lesser flow discharge and a straighter direction when compared to the upper section. On the other hand, the geomorphological landscape of the area is characterized by the presence of alluvial high and low terraces. The morphogenesis and development of the terraces are associated with the evolution of the river during the recent Quaternary (e.g., Iriondo 1996; Journaux and Pellerin 1980; Panario and Gutiérrez 1999).

Early human occupation evidence has been recovered in situ at the basal deposits of the lower terrace. On the other hand, the evidence for middle and late Holocene archaeological record is scattered over a wider area, comprising the upper terrace, river banks, and other recent alluvial deposits in surface and subsurface layers of such profiles (e.g., Capdepont 2012; Castiñeira 2008; Castro 2017; Castro and Costa Angrizani 2014; Costa Angrizani 2012). The most noticeable recognition of the characteristics and distribution of the archaeological record of the area corresponding to the results obtained by Salto Grande Archaeological Rescue Mission during the 1970s and 1980s (Guidón 1984, 1989). The archaeological rescue activities were developed in order to gather all possible evidence before the area was flooded by the Salto Grande hydroelectric dam. Currently, this dam supplies energy to both Argentina and Uruguay.

Nearly 35 years after the Salto Grande Archaeological Rescue Mission, it has been proposed to update the archaeological knowledge of the coastal area of the MBUR. The main goal of this work is to contribute to the knowledge of the sociocultural evolution of the littoral areas of the La Plata River basin. This aim is integrated into the recognition of the pre-Hispanic human strategies developed in areas of high hydrometeorological vulnerability (Castiñeira et al. 2017). The methodology chosen in this contribution has an interdisciplinary nature and it is inscribed in the framework of an alluvial geoarchaeological investigation (Ferring 2017).

The chronicles from the sixteenth century, during the early stages of European colonization, identified a multiethnic confluence in our study area. In these documents, the existence of strong degrees of interaction between social groups based on, among other strategies, the maintenance of alliances, conflicts, kinship relations, and exchange of technological products is highlighted (Apolinaire 2016; Apolinaire and Bastourre 2016; Capdepont et al. 2017). The archaeological evidence recovered in this research allows to consolidate the characterization of the area as a "support-area" of production, distribution, and circulation of cultural materials. Moreover, the sedimentological and chronological results obtained here contribute to the knowledge of the palaeoclimatic and paleoenvironmental characteristics that have dominated the studied area during the recent Quaternary in the fluvial littoral of the Paraná-Uruguay basin.

Characteristics of the Study Area in the Present and in the Past

The study area, which is located in the MBUR (Fig. 1), presents an exuberant and original vegetation of subtropical influence. Typical marginal forests are developed in the coastal zone of the Uruguay River and on the margins of its tributaries. This type of vegetation characterizes the subtropical phytogeographic province described by Cabrera (1953) and to which he named as "Galería uruguayense". At the zoological level, this area represents the southernmost distribution of subtropical lineage (see Arzamedia et al. 2015). According to paleontological studies of the area, the present faunal configuration began to be recognized during the late Pleistocene (e.g., Ubilla 1996; Ubilla and Perea 1999). The climatic conditions that marked the end of the Pleistocene probably favored the development of riverine forests and the colonization of the Brazilian fauna in the area (Ubilla et al. 2004). The fossil record allows arguing that the landscapes of the MBUR were colonized by mammals adapted to fluvial and lacustrine environments associated with riparian vegetation (Goso and Ubilla 2004).

For the early Holocene, according to the sedimentological, biocompositional, and chronological proxies analyzed by Castiñeira (2008), the climate has gradually evolved to temperate subhumid conditions with moderately seasonal rainfall. At present, the Uruguay River Basin has a humid subtropical climate with an average annual temperature ranging from 19 to 21 °C and an annual average rainfall ranging from 1,600 mm in the southwest to 2,100 mm in the northeast. The current climatic-environmental conditions may have begun to develop around ca. 2,500 ^{14}C years BP (Capdepont 2012).

From an external geomorphological point of view, the evolution of the recent Quaternary in the MBUR is characterized by the construction and development of the high and the low terraces (Fig. 2). The higher terrace would have begun to be constructed during the late Pleistocene; whereas the lower terrace, shortly before the beginning of the Holocene, ca. 13,000 ^{14}C years BP (Iriondo 1996; Iriondo and Kröling 2004, 2008; Panario and Gutiérrez 1999). On the western margin of the basin, the deposits of the lower terrace constitute the Concordia Formation (Iriondo and Kröling 2004, see Figs. 1 and 2). The Concordia Formation has two sedimentary members: Lower and Upper (Iriondo and Kröling 2004). Both members are the result of paleo-levees juxtaposition; furthermore, several pedogenetic processes are recognized in their sequences. The basal levels of the Lower Member may have been deposited during the semiarid climatic conditions that characterized the end of the Pleistocene. During this time, large sporadic floods generated the depositional and the accretion of their fine sediments (Iriondo 1996). On the other hand, the development of paleosols characterized the cuspidal deposits of the Lower Member. Its genesis and evolution are associated with the conditions that dominated during the Hypsithermal period (Middle Holocene). Finally, the Upper Member may have started depositing during the Late Holocene, and it is characterized by deposits of flood facies and by the sectorized presence of soil development. The latter is correlated with the conditions

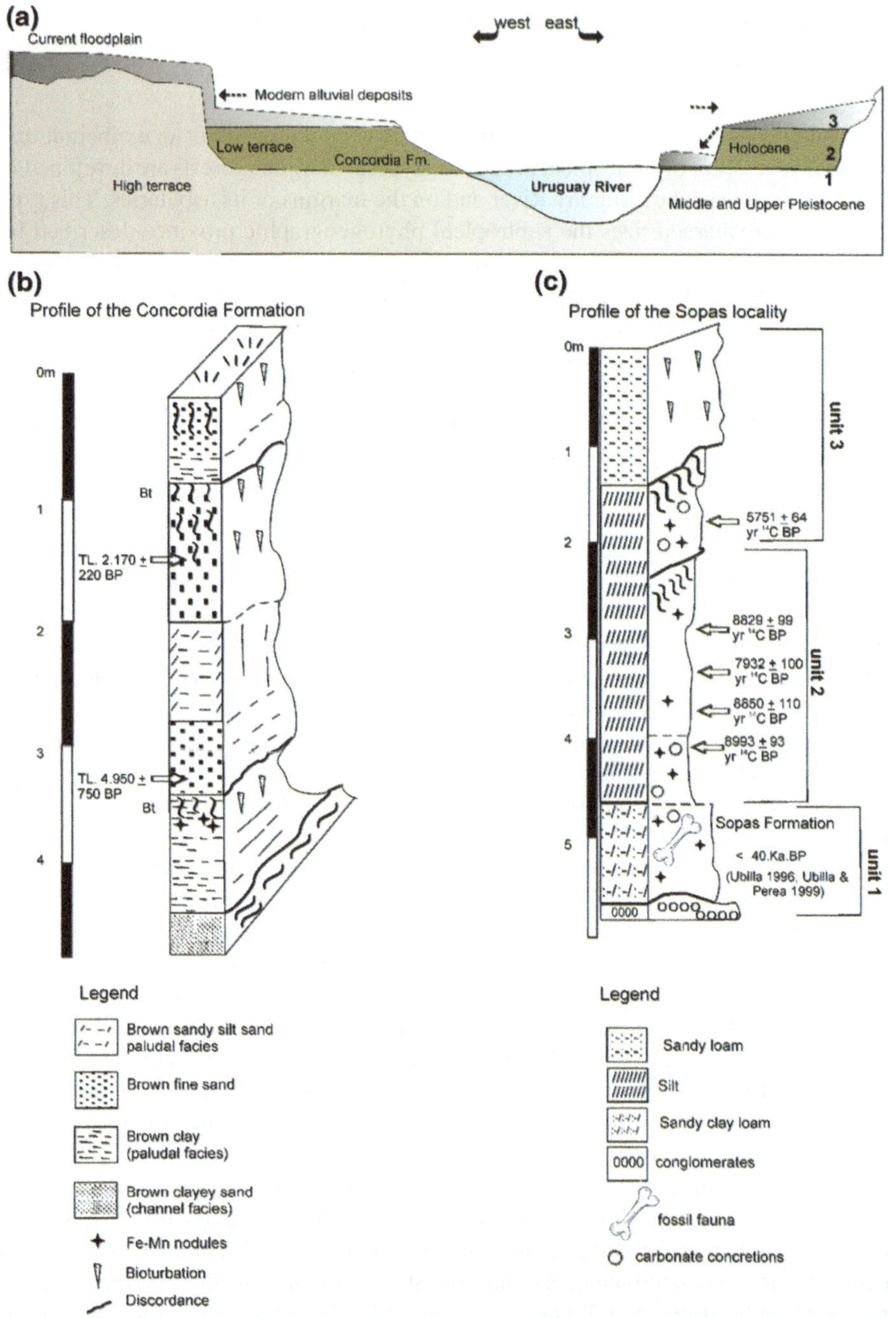

Fig. 2 **a** Scheme of the fluvial terraces of the MBUR. **b** Stratigraphic profile of the Concordia formation (taken from Iriondo and Kröling 2008). **c** Stratigraphic profile of the Sopas locality studied by Castiñeira (2008) and the radiocarbon data obtained for Units 2 and 3 remained unpublished (see Table 1)

that prevailed during the Medieval Warm Period ca. 1,400–800 ^{14}C years BP (Iriondo and Kröling 2008).

The process of human occupation in the landscape of the MBUR appears synchronously with the deposition of the lower terrace. The radiocarbon dates obtained by Suárez (2000, 2003) and Suárez et al. (2017) for the Pay Paso and the Tiger archaeological sites confirm the Pleisto-Holocene age of this process. It has been hypothesized that for the success in the human colonization it was necessary to maintain the social connectivity in a wide-scale range (Castiñeira 2008). The latter is supported by the presence of technologic features shared among the early lithic industries of Uruguay, southern Brazil, and the Pampas region of Argentina (Castiñeira et al. 2012; Flegenheimer et al. 2003; Hilbert 1991; Nami 1994, 2007; Politis 1991; Suárez 2004).

According to the results obtained by the Salto Grande Archaeological Mission Rescue for the Middle Holocene, ca. 5,000 years BP, the number of hunter-gatherers-fishermen groups had considerably increased in the area by that time (Guidón 1989). This statement was based on the recognition of a greater number of sites assigned to the period and the wide presence of debris and tools, such as polished and carved instruments. Moreover, Capdepont (2012) inferred a population increase as responsible for the human development under semiarid–arid climatic conditions.

For the late Holocene, the archaeological literature that refers to the middle basin of the Uruguay River recognizes several cultural entities: "Enterriana" (also called "Básica del Litoral"), "Ribereños Plásticos" (or "Goya – Malabrigo"), "Tupiguaraní" and "Guaraní" (Capdepont 2012; Costa Angrizani 2012). The "Entrerriana" culture developed around 2,500 years BP, according to ^{14}C dates obtained for the sites located in the region of the islands of the Uruguay River (Baeza et al. 1977; Hilbert 1991). The settlement pattern of this entity corresponds to the occupation of elevated areas on the margins of rivers and swamps. The cultural material generated by these fisherman and hunter groups is characterized by pieces of pottery, carved, and polished tools, artifacts in bones and mollusk shells (e.g., Hilbert 1991; Serrano 1972). In its pottery pieces, the dotted and zig-zag decorative motifs stand out (Serrano 1972).

Approximately at 2,000 ^{14}C years BP, the registry assigned to the "Ribereños Plásticos" was added to the archaeology of the area. This entity, also known as the "Goya-Malabrigo" (e.g., Cerutti 1992, 2000; Politis and Bonomo 2012) was defined by morphological and stylistic traits of pottery, specialized technologies for the exploitation of aquatic resources, small-scale farming, and building of mounds or "cerritos" (Politis et al. 2011; Castiñeira et al. 2013). For Cerutti and González (2007), "Goya – Malabrigo" represents the maximum expression of the use of the fluvial system. The occupation pattern included a wide network of settlements, which constituted an effective sociocultural response to the risks involved in the colonization of highly dynamic wetlands (Castiñeira et al. 2017). Subsequently, and around 1,000 ^{14}C years BP, the study area would be the scenario of the dispersion of the "Guarani" tradition. The evidence for that widely distributed in superficial sites. The diagnostic elements assigned to this tradition are brushed thick walled ceramics with polychrome paint, corrugated and ungulate decoration, and polished axe heads (Bonomo et al. 2015; Castro and Costa Angrizani 2014; Costa Angrizani 2012). The

"Guarani" settlement pattern is circumscribed to those territories in the fluvial littoral zones. Their agricultural practices were frequently associated with the presence of extensive patches of anthropogenic dark earth. The practice of secondary burials in urns can be added to the description (Constantine 2014; Noelli 2004). According to the investigations of Costa Angrizani (2012), the axes of the Guaraní spatiality on the coast of the Uruguay River would have been strategically delimited by the fishing, hunting, gathering, supply of lithic and sedimentary material for pottery and by the area of arable land in the riverine forests. The sites located in the alluvial plains can be assigned to residential contexts. The cultural remains found in these sites can be associated with several human activities such as the manufacture of instruments, cultivation and food processing. Finally, the sites distributed in the highest areas of the fluvial landscape (for example, hillocks and tops the hilltops) are mainly characterized as workshops for collecting lithic raw materials and temporary camps, possibly linked to the exploitation of resources in the transition zone between the forest and the plains.

In order to contribute to the knowledge of the evolution of the climatic, environmental, and cultural characteristics in the MBUR during the Holocene, this project began to implement alluvial geoarchaeological investigations. The study of alluvial systems and their natural and cultural records is very important in archaeological surveys (Ferring 2017).

Geoarchaeological Background for the Study Area

When it comes to recreating the interactions between humans and nature, geoarchaeology constitutes the methodological key for archaeologists (e.g., Hill 2017). In addition, geoarchaeology has been an essential tool when framing, with more precision, the temporal scale of the cultural records preserved in fluvial environments (Gilbert et al. 2017). Both applications have marked the history of archaeology in the basin of the Uruguay River and have generated the background for the development of the current alluvial geoarchaeology investigation (Ferring 2017).

Since the 1970s, the recreation of the paleoecological scenarios, in order to contextualize the sociocultural development, has been one of the main objectives of the archaeological investigations of MBUR (e.g., Bombín 1975, 1976; Bórmida 1964; Cigliano 1968; Lafón 1971; Rodríguez and Rodríguez 1984; Serrano 1932, 1950, 1972; Taddei 1964). A fundamental issue was the need to obtain relative geochronologies due to economic and logistical difficulties related to numerical dating techniques. Different relative dating methods were used in order to generate the cultural schemes up to the 1970s. For example, some chronologies were inferred using the diagnostic value of some products or decorative styles, whereas others, by the presence or absence of a distinctive character that represented a stage in the sociocultural evolution; the cultural industries were granted with different time frames based, in some cases, on the hypothesis that on the hypothesis that simpler artifact forms are earlier than more complex objects. The relative chronologies of the objects could also be obtained by the relative position of the strata in the geological unit, without considering that the redeposition of materials and erosive processes that occurred in the areas could have been affected by fluvial dynamics. Finally, the correlation using chronostratigraphic schemes generated for other better-studied areas of the continent was also used in order to get the relative dating. An example of the latter is the evolutionary paleoecological model proposed by Bombin (1976) for the late Quaternary of western Rio Grande do Sul—Brazil, Campanha region (Touro Passo Formation; Fig. 1), which was widely used by other researchers. This model was based on the lithological changes of the alluvial deposits associated to the American glacial model (Flint 1971), thus adjusting the Touro Passo Formation paleocultural evidence of the Paleoindian paradigm to the climatic changes proposed for North America (Haynes 1964; Krieger 1964; Miller 1976, 1987).

The Touro Passo Formation has been the lithostratigraphic unit used to synthesize the environmental and sociocultural climatic evolution during upper Pleistocene—Holocene time of the Uruguay River basin (e.g., Antón 1975; Bombín 1976; Miller 1987). The results were correlated with those obtained for the Sopas Formation (Northern Uruguay, Antón 1975) and with those from the Luján Formation (Buenos Aires Province, Argentina; Fidalgo et al. 1973). For instance, the layer of volcanic ashes identified in the Touro Passo sedimentary succession, with a chronology of 10,000 years BP according to Bombin (1976), was considered as a "chronological marker" for Paleoindian research of the region (Antón 1975; Miller 1987; Guidón 1989). In addition, the record of Pleistocene mega-mammals, assigned to the basal deposits of the Touro Passo Formation, were used to consolidate the Paleoindian period in the basin of the Uruguay River. However, the association of archaeological materials with remains of the extinct Pleistocene fauna in the basin of the Uruguay River is controversial (in this regard, see Castiñeira 2008, 2009; Días Schmidt 2004; Pouey 2018). The estimated age for the "chronological marker" layer was originated from another chronological correlation. This estimation comes from the work of Heusser (1974), who proposed a chronology of 10,000 years BP for the deposits of volcanic ash at the southern lake district of Chile. Such age was chosen for the volcanic ashes of Touro Passo Formation. In 1990, according to new ^{14}C chronologies and geochemical results, different ash emissions, generated by the Andean volcanoes during the Late Pleistocene and Holocene, were recognized (e.g. Stern 1990, 1991, 1992). These results allowed questioning the chronological estimation of Heusser (1974) and its derivations. In addition, the development of tephrochronological investigations in the fluvial deposits on the eastern margin of the MBUR provided chronologies by ^{14}C of around 4,000 and 3,800 years BP (Bracco et al. 2001; Castiñeira and Fernicola 2005). These data reinforce the questioning of the proposed chronological inferences for the Touro Passo Formation, as well as for the geological correlations proposed in the case of the Sopas Formation (Antón 1975) from northern Uruguay. Therefore, it is necessary to obtain more precise data in order to increase the knowledge regarding the paleoecological evolution of the MBUR.

Methodology

This work presents and discusses the recent results achieved in those studies developed in the western margin ca. 31°16′ and 31°40′ south latitude and 58°20′ west longitude (Fig. 1). In 2016 an extensive fieldwork was carried out, aimed to sketch land and fluvial possible mobility routes and detect archaeological sites along the Uruguay River. During this fieldwork, we select different coastal sedimentary successions according to their degree of resolution. In this paper, the results achieved in the selected stratigraphic successions: H2SCU1 and C1SECU2 were presented. In addition, it is herein included the recent chronological data for the sedimentary sequence T2ASIV obtained for the Sopas locality (31°23′30.4″S/56°45′37.95″W, Fig. 1), which is located at the eastern margin the MBUR. This locality was studied by Castiñeira (2008), but only a numerical reference date was available for the palaeoenvironmental and paleoclimatic inferences proposed by the author (Table 1 and Fig. 2). Therefore, the inclusion of new dates for the Sopas locality allowed discussing the evolution of alluvial deposition in the area during the Holocene.

The succession H2SCU1 is located at 31°31′42.3″S and 58°00′57.9″W, and it has 7 m in thickness. The exposure of this sedimentary sequence is the result of the recent anthropogenic construction of a drainage channel (Fig. 3). Archaeologically, H2SECU1 is characterized by the presence of lithic and ceramic materials that are concentrated and distributed in the superficial deposits of the bed of the channel. The archaeological potential of this site, called "Paso del Hervidero", was previously recognized by Rodríguez and Weissel (2013).

The succession C1SECU2 is located at 31°32′46″S and 58°00′19″W at the coastline of the Uruguay River about 2 km upstream of the H2SCU1 sequence. The cutbank has a vertical expression of approximately 1.30 m. The stratigraphic sequence is characterized by the presence of a paleosoil (Fig. 3), which has a lateral continuity along more than 12 km and it presents, in some sectors, carbonized woody material. At the beach zone, associated with the stream bank, two pottery sherds were recovered.

In the successions H2SCU1 and C1SECU2, stratigraphic surveys and sediment sampling were carried out. The archaeological materials contained in the stratigraphic deposits and those present in the superficial deposits were collected for laboratory analysis. Coals were also collected for specific characterizations and dating. Radiocarbon dating was performed by the Laboratorio de Radiocarbono (LATYR)-CIG, CONICET-UNLP (La Plata, Argentina) by means of benzene synthesis and the liquid scintillation counting (LSC) method (Huarte and Figini 1988) with PerkinElmer TriCarb 3170 TR/SL and Packard TriCarb 1050 TR/LL spectrometers. Ages are expressed in conventional radiocarbon years before present (1950 AD) and corrected by isotopic fractionation with $\delta^{13}C$ values estimated by the table (Stuiver and Polach 1977). The dates were calibrated with CALIB 7.0.4 using ShCal13 ^{14}C calibration curve (Hogg et al. 2013).

The stratigraphic survey was carried out by means of the perceptive discrimination of changes in the sediment texture, the presence of sedimentary structures, pedological features of the deposits, biological pertubation colorimetric aspects

Table 1 Radiocarbon dates generated by this research for H2SECU1, C1SECU1, and Sopas locality

Site	Deep (m)	Dated material	Lab. code	Date ^{14}C (yr BP)	Date cal BP (range 1σ)	References
Sopas locality 31°23′30.4″S/56°45′37.95″W	4	Sediment	Erl-13594	8,993 ± 93	[9919 : 10221]	In this work
	3.50	Sediment	Erl-13593	7,932 ± 100	[8590 : 8979]	
	2.90	Sediment	Erl- 13592	8,829 ± 99	[9614 : 10109]	
	2.40	Sediment	URU 0431	8,850 ± 100	[9682 : 10135]	Castiñeira (2008)
	1.30	Sediment	Erl- 13591	5,751 ± 64	[6410 : 6601]	In this work
H2SECU1 31°31′42.3″S/58°00′57.9″W	2	Sediment	LP- 3441	950 ± 50	[761 : 905]	In this work
C1SECU1 31°32′35,40″S/58°0′23,41″W	1.20	Charcoal	LP-3434	1,550 ± 60	[1315 : 1466]	
C1SECU2 31°32′46″S/58°00′19″W	1.40	Sediment	LP- 3450	1,840 ± 60	[1615 : 1821]	

Fig. 3 **a** "Paso del Hervidero" locality (Google Inc., image date: 4/5/2017, last accessed: 10/19/2018). **b** H2SECU1 succession. Detail of the paleosoil that contains archaeological material in situ. **c** C1SECU2 succession. Detail of the paleosoil (U2)

(Munsell color chart), and presence–absence of archaeological evidence and faunal remains. Sediment samples were submitted to textural, compositional (mineral, micro-archaeological and biosiliceous contents) and chronological analysis. Textural analyses were performed according to standard methodologies (Carver 1971; Day 1965). The grain size classification was made following the parameters proposed by Folk (1954).

The composition of sized-samples retained in sieves (2–0.062 mm) was studied with binocular magnifying glasses in order to identify siliciclastic, and bioclastic remains, and micro-archaeological materials. The fine to very fine sand fraction (0.125 mm) was studied with polarization microscopy for the mineral composition analysis. In the silt fraction (0.05–0.005 mm), the biogenic silica was observed through optical microscopy following the same techniques and criteria detailed in del Puerto (2011), del Puerto et al. (2016), Zucol and Colobig (2014).

The lithic artifacts were analyzed in their techno-morphological aspects following the procedures proposed by different authors (Andrefsky 1998; Aschero and Hocsman 2004; Dias Schmidt and Hoeltz 1997; Shott 2003). The analysis of potsherds was carried out following the suggestions of Balfet et al. (1992), Capdepont (2012), Cremonte and Bulgliani (2009), and Hally (1983). Finally, the ceramic pastes characterization analyses were carried out through polarization microscopy, following the procedures described by Orton et al. (1993) and Roper et al. (2010).

Results

"Paso del Hervidero" H2SECU1 succession consists in 11 strata that were grouped in three different sections (Fig. 4). The basal section of the succession of ~2 m thick develops over gray (10 YR 5/1) sandy mud deposits (E11) without an exposed base. This section is also made up of muddy sand deposits (E9 and 7) in which sandy lenses (E10) and gravelly sand lenses (E7) are intercalated. The deposits of the basal section did not present archaeological evidence.

The middle section was developed overlying and in erosion contact with the basal section. It has a thickness of ~3 m and pedological aggregation structures are present

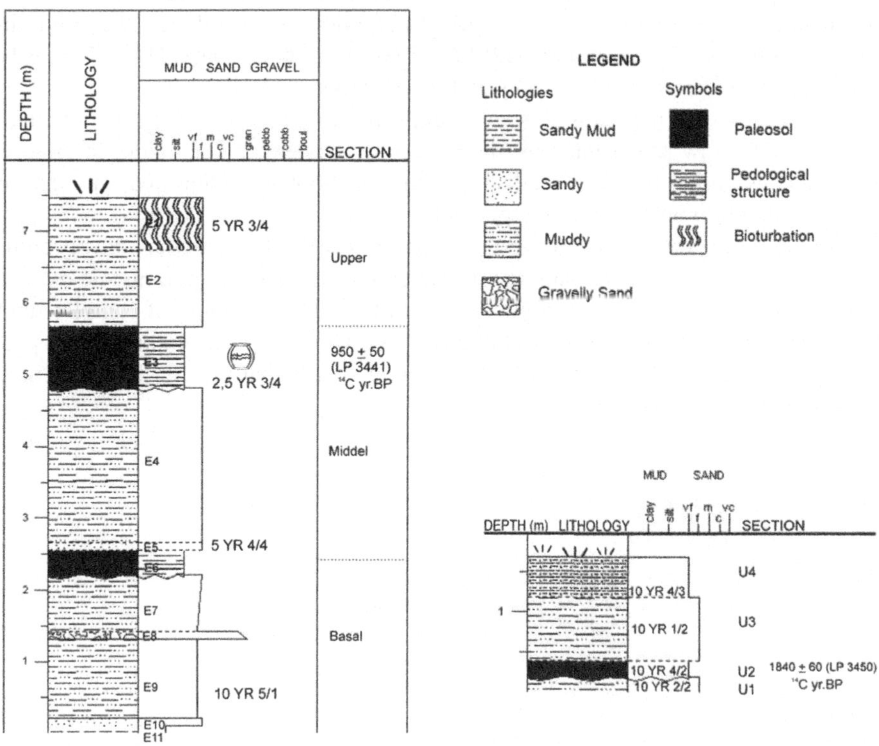

Fig. 4 Left: H2SECU1 stratigraphic profile showing sections and radiocarbon date. Right and down: C1SECU2 profile showing units and radiocarbon date

(e.g., blocks in E6 and E3). It is characterized by dark sandy mud levels (2.5 YR 3/2) and dark reddish muddy sandy levels (5 YR 3/3). These levels are separated by the sandy lens (E5) of reddish-brown color (5 YR 4/4). Finally, the top section has a vertical development of ~1.80 m and it is characterized by the presence of muddy deposits (E 2.2 to 1) of brown color (5 YR 3/4); it constitutes the current soil covered by vegetation.

The composition of the samples from the succession H2SCU1 was characterized by the domain of the following minerals: quartz, plagioclase, chalcedony, K-feldspar (sanidine), micas (chlorite, muscovite and biotite), pyroxene, and opaque. The presence of volcanic glass was also recorded in the observations. The biosiliceous content is mainly represented by phytoliths, sponge spicules, and diatom frustules.

The cultural evidence in situ were only found in stratum E3, which is the top of the second section of the H2SECU1 succession. The organic matter content of E3 was dated by the conventional method of ^{14}C, and an age of 950 ± 50 years BP (LP3441) was obtained (see Table 1).

The pottery sherds recovered at the site "Paso del Hervidero" (N = 31) showed variability regarding the preservation states in line with their taphonomic context. Evidence of wear, roundness, and eroded surfaces with detachment and/or exposure of the temper material dominate the sherds that have been recovered in the bed of the channel (superficial context, N = 14). In addition, the sherds recovered within the E3 stratum of the successions H2SECU1 (stratigraphic context N = 17) mostly show evidence of smoothed surfaces and roots marks.

In general terms, the ceramic sherds recovered in both contexts have smoothed surfaces and a low representation of surfaces with engobe. The colors of engobe of the pieces varied between reddish and brown. Soot was identified on external surfaces, although with very low representation.

By thin sections analysis, it was possible to observe that the pastes are mostly composed of fine material with abundant quartz clasts, feldspar concretions, Fe oxides, and few plagioclase and isolated micas. In addition, a high content of complete and fragmented sponge spicules was also observed. Moreover, the temper in the pottery is characterized by the presence of ground sherds that reach a fine sand size (250–125 μm) to coarse silt (63–31 μm).

Table 2 Diversity of lithic products recovered at the "Paso del Hervidero"

	Raw material						
Typological classes	Sandstones	Basalts	Chalcedony	Silicified limestones	Other Silicificat.	Undet.	Total
Flint	57	9	35	13	17	0	131
Cores	3	3	1	2	0	0	9
Instruments	2	0	0	0	1	0	3
Geofacts	10	0	6	0	0	3	19
Total	72	12	42	15	18	3	162

The raw materials that characterize the lithic material recovered at the "Paso del Hervidero" locality correspond to siliceous rocks (chalcedony) and sandstone with a representation of between 46.3% and 44.4%, respectively (Table 2). Unipolar direct percussion was the technique most used for the core reduction and flake production. In the sample analysis, evidence of several stages in the sequence of exploitation of the rolled pebbles has been observed: elimination of the external crust of the pebbles, extraction of flakes, and manufacture of instruments. The three recovered instruments were manufactured from flakes, and are bifacial blade instruments with angles between 45° and 60°.

Seven strata gathered in four lithostratigraphic sections were differentiated in the succession C1SECU2 (Fig. 4) in the coastal line sector. The basal section (U1) of ~0.20 m thick, without exposed base, is characterized by brownish black color (10YR2/2) and sandy mud deposits. In net contact whit U1 a second section (U2) of a thickness of ~0.20 m develops. This U2 was characterized by the presence of a paleosol which is visible in the coastal strip for approximately 12 km between the mouth of the Arrebata Capa stream and the mouth of the Yerúa stream (Fig. 1). By sectors, the paleosol can reach 0.40 m of vertical expression. It is identified by its black color (10YR 1/2) and by mud to sandy mud texture. The radiocarbon age obtained from the organic content of the sample of this paleosol was 1,840 ± 60 years BP (LP 3450) (Table 1). In gradual contact above U2, a new thick unit (U3) is developed of 0.80 m. It is characterized by grayish brown color (10 YR 4/2) and a muddy sand deposit. The successions C1SECU2 culminates with a fourth unit (U4). The U4 unit develops gradually over U3 and is characterized by a muddy sand to sandy mud deposit. It has an average thickness of 0.50 m of brown color (10 YR 4/3). This last unit supports the current vegetation cover.

The mineralogical composition of the samples of C1SECU2 succession was not different from the composition observed in the samples of H2SCU1 succession. However, the samples of the basal sections exhibited a greater presence of chlorite, muscovite and biotite.

On the other hand, the preliminary biosiliceous analysis for the samples obtained from C1SECU2 succession allows distinguishing an association characterized by the presence of abundant sponge spicules, grass phytoliths (Podostemaceae + Poaceae + Cyperaceae + Dicotyledons) and diatoms.

A few 300 m upstream of C1SECU2, and in stratigraphic association with the already described paleosol, samples of a concentration of charred woody material were obtained. This area of coals concentration was cataloged as C1SECU1 (31°31′42.3″S/58°00′57.9″W). Currently, the nature of this concentration has not been determined, but a radiocarbon age was obtained for a subsample of the set of found coals. The obtained radiocarbon age was of 1,550 ± 60 years ^{14}C BP (LP 3434 in Table 1).

In none of the deposits of C1SECU1 sequence archaeological materials were found. Nevertheless, two pottery sherds were recovered in spatial proximity with C1SECU2 and C1SECU1 on the coastline between the mouths of the Yerúa and Arrebatacapa streams. Both pottery sherds correspond to edge sectors. One is characterized by the presence of smooth surfaces with adhesions and a rounded fracture

Fig. 5 Cultural materials recovered in this research. **a** Pottery sherds analyzed from "Paso del Hervidero" locality; **b** photomicrographs of thin section analysis in pottery sherds of the "Paso del Hervidero" locality. Observe the presence of ground sherds (size 175 μm). **c** Lithic material of the "Paso del Hervidero" locality

surface. The other pottery sherd (C13) have external surface an incised motif of triangles aligned in horizontal sequence parallel to the edge, in which the bases constitute a continuous straight line (Fig. 5).

In the sequence T2ASIV of the Sopas locality, Castiñeira (2008) discriminated three depositional units (Fig. 2). The basal unit (Unit 1) develops above ground deposits and on the basaltic basement; it is constituted by alluvial deposits that exhibit fine horizontal stratification associated with primary current structures. The deposits of this unit include fossil remains of Pleistocene fauna and were correlated with the deposits of the Sopas Formation defined by Antón (1975) and chronologically positioned in the interval that characterized the Marine Isotopic Stage 3 development in the region (see Blasi et al. 2016; Ubilla et al. 2016). Above Unit 1, Unit 2 was developed, which was characterized by the development of a paleosol of 1 m of power (see Fig. 2). Until very recently, Unit 2 had a unique radiocarbon dating of 8,850 ± 100 years ^{14}C BP (URU 0431 in Table 1). Three recently reported radiocarbon ages (see Table 1) confirm the chronology assigned by Castiñeira (2008) for the development of the unit during the early Holocene. Sequence T2ASIV ends with Unit 3 on whose top is the current alluvial soil. This unit has not been dated; however, its depositional genesis and pedogenetic modification had been assigned to the Middle Holocene. The recent Erl-13591 dating (see Table 1) which indicates an age of 5,751 ± 64 years ^{14}C BP confirmed the previous chronological estimation (see Castiñeira et al. 2011).

Discussion and Conclusions

The results of the archaeological investigations developed in the littoral of the MBUR constituted an important link in the knowledge of the continental prehistory and in particular of the South American lowlands. It contributed to the knowledge of the antiquity of the human population process, its dispersion and the sociocultural characteristics of its inhabitants (e.g., López 2013, Politis et al. 2015). For the Holocene archaeology of the South American lowlands, these researches also contributed to the recognition of the dispersion of Amazonian populations, the development of ceramic technology, the understanding of the emergence of social complexity, adoption of crops and processes of inter-ethnic social interaction (e.g., Apolinaire 2016; Bonomo et al 2015; Noelli 2004). In this sense, the records contained in the alluvial deposits of the littoral area of the MBUR allowed to conceive the pre-colonial archaeology of the region as a long-term history. The understanding of the depositary evolution linked to fluvial dynamics gives us the possibility of reconstructing the temporally extensive and diachronically slow historical processes (such as population expansion, territorialization, interaction, migration, occupation, exploration, and transformation of the environment) that delineate the cultural history of the people.

The studied alluvial sequences are records that show the recurrence of erosional and depositional episodes that are characterized by the fluvial dynamics, during the Holocene of the middle Uruguay River in the study area. This dynamic is recorded in the sequence studied by Castiñeira (2008) for the eastern margin of the river and is documented in the new reported chronologies data for the Early Holocene paleosols (code Erl 13594, 13593, 13592 in Table 1), and for the Middle Holocene paleosol in the Sopas locality (code Erl 13591 in Table 1), whereas, for the western margin, this recurrence of events during the late Holocene has a clear expression in the sequences H2SECU1 and H2SECU2.

The erosive and depositional episodes recorded can be linked to the evolution of paleoclimatological phenomena supported by regional models such as climatic changes on a secular scale associated, for example, with ocean circulation patterns, variations in solar activity, among other phenomena that affect climatic variability (e.g., del Puerto et al. 2011; Favier Dubois and Zárate 2012; Laprida et al. 2009; Piovano et al. 2009). In turn, the records studied here allowed to recognize the recurrence of different hydrometeorological events at the basin scale that affected the study area. Inferences of paleoclimatic and paleoenvironmental dynamics allowed to evaluate if the intensity, recurrence and/or permanence of hydrometeorological events were considered in the pre-Hispanic strategies of landscape selection. In addition, the obtained results allowed to evaluate the incidence that the fluvial dynamics has in the preservation of the archaeological record. Regarding this last point, the state of the surface of the pottery sherds, which have been recovered in the bed of the channel of the site H2SECU1 and in the coastline associated with the sequence C1, shows the effects generated by the removal and the hydrological dragging. Whereas the state of the surface of the sherds, that have been recovered in stratum E3 of the successions H2SECU1, presented evidence of biostratinomical (root marks) and diagenetic alterations (superficial detachments) that affected the pedological contexts with greater intensity.

The paleoenvironmental evolution of the region and the hydrometeorological dynamics of the area are recorded in the biosiliceus associations contained in the studied alluvial deposits. Following the compositional preliminary studies, the association present in the basal deposits of the C1SECU2 successions correlates with high-energy fluvial processes that have favored the development of high water (flood) phases that have affected the coastal space. The phytolith assemblages of these samples can be associated with the development of the communities that characterize gallery forests. The presence of morphotypes attributable to elements of subtropical–tropical flora would allow establishing that the climatic conditions were much more humid and warmer than the current ones. The stabilization of this climatic phase of higher humidity and heat favored the development of the paleosol that characterized the U2 and that extends for approximately 12 km in the coastal levee. Subsequently, in the cuspidal deposits of the sequence, it was preliminarily recognized a greater dominion of graminoid elements of open spaces and a decrease of elements associated with the development of the "Uruguayan Gallery" in the bioproxies contained.

On the other hand, the sedimentological, chronological and archaeological results obtained here allowed to correlate the deposits of the studied sequences with the constituent deposits of the Upper Member of the Concordia Formation defined by Iriondo and Kröling (2004). The Upper Member of the Concordia Formation is characterized by deposits of flood facies and pedofacies that have developed during the late Holocene. The Upper Member is correlated in this paper with the U1 of the successions C1SECU2 and with the basal deposits of H2SECU1. In this sense, the presence of very fine pebbles (from 2 to 4 mm) is added in strata E8 and E10 of the basal section of H2SECU1 to the paleobotanical evidence that refers to phases of high waters.

During overflow events (flood facies), the solid suspension load (silt and clay sediments) are deposited in the floodplain and in the geoforms that cut to the floodplain. To a lesser extent, psamitic and psefitic clastic materials are also deposited; these could have mobilized as bed load by fluvial currents or have been mobilized from the margins by fluvial coastal waves. Likewise, and in relation to the hydrometeorological events, the presence of pebbles and archaeological materials eroded in the costal line, in the bed of the channel, or included in the deposits of the low terrace, may be a consequence of erosive processes that have affected the highest areas of the landscape and have reached the depressed areas by slidings.

Both studied sequences (C1SECU2 and H2SECU1) are the result of the deposition of fine sediments (mud, sandy mud, muddy sand) by low-energy overflow. These alluvial accretionary records have been in turn modified by pedogenesis. The pedogenetic modification suggests the existence of intervals in the processes of alluvial deposition and/or a decrease in their rate of accumulation. Thus, the alluvial deposits are constituted as the parental material that has supported the vegetal communities in the past and that, at the top of both studied sequences, sustained the current vegetation. The mineralogical composition of the sand fraction is genetically correlated with the volcaniclastic contributions. Particles generated by erosion of the basalts are transported in suspension by the Uruguay River. As suggested by the presence of fresh volcanic glass, there is as well participation of distal eolian contributions, mostly located in the Andes.

According to these results, it should be considered the evolution of two pedological cycles or phases of morphogenetic stability in our study area during the period in which the Upper Member of the Concordia Formation was deposited. A first cycle would be represented in the U2 of C1SECU2 sequence with a minimum reference age of ca. 1,800 ^{14}C years BP (LP 3450; see Table 1). The concentration of charcoals which dating was ca. 1,500 ^{14}C years BP (LP 3434; see Table 1) supports this first recorded geomorphological stability process. The remains of coals correspond to the natural or anthropic combustion of plant species that colonized in the past the paleosurface of the area. The pedogenetic development in the study area was associated with the warmer and more humid climatic conditions that characterized the beginnings of the late Holocene in the region (e.g., del Puerto 2011; Quattrocchio et al. 2008; Tonello and Prieto 2010). Future chronological analyses will allow to discuss the integration of the paleosol recorded as E6 in the successions H2SCU1

The second cycle would be represented by the uppermost deposits of the middle portion of the successions H2SCU1 (E3). The LP3441 dating (ca. 1,000 ^{14}C years BP, see Table 1) provides a minimum reference age for the latter pedogenetic process. It also allowed to suggest a temporary position of the human occupation that has generated the material recovered in the "Paso del Hervidero" locality.

It is proposed that the human occupation, which was recorded in "Paso del Hervidero", should be contextualized with the development of palaeoclimatic conditions recognized for the interval widely recognized as the Medieval Warm Period (800–1200 AD). During this warm period, there would have been an expansion of the river systems (Apolinaire 2016) in the study region. This expansion, following what was stated in Apolinaire (2016), would have facilitated the mobility of the human groups by land and inland waterways for the provision of various resources and for the maintenance of social networks. It would have also favored the spatial and topographic location of the settlements and other areas of strategic value.

The coastal location of the "Paso del Hervidero" site and its cultural record, which is characterized by the use of artifacts, the supplying of raw materials, the production of ceramics and lithic tools, allowed to assign it to the settlement pattern described by Capdepont (2012) for the residential sites along the Uruguay River during the Medieval Warm Period.

Each lithic material recovered at the "Paso del Hervidero" site-western margin—-can be associated to the steps involved in the operational chain in the lithic technology. The evidence suggests that the supply of raw materials, its transformation, usage, and subsequent abandonment took place in these sites. On the other hand, the registered morphotechnological characters allowed to infer a low input work in the manufacture of the recovered lithic materials. The development of an expeditious technology may be in line with the wide and varied supply of mineral resources that are available in the area (Apolinaire 2018). Given the characteristics of the residual crust observed in the pieces, it may be established that the supply of the lithic raw materials was carried out mainly in the secondary deposits of the river coast and to a lesser extent in rocky outcrops (basalts and sandstones).

The ceramic technology recovered in the primary and secondary contexts intervened by this research share techno-morphological characteristics with the materials recovered by Capdepont (2012) in the late Holocene pottery sites located on the eastern margin of the Uruguay River. An example of this is the identification of ground sherds, concretions of oxides of Fe and spicules of sponges in the paste in the pottery sherds analyzed here. Likewise, the raw materials and techno-morphological characteristics of the lithic industry that have been herein studied are similar to those recognized by Apolinaire (2016) in the sites located in the inner plain of the western margin of the middle basin of the Uruguay River.

The cultural records present in the study area supports what was proposed by Capdepont (2012) and Apolinaire (2016) regarding the existence of circulation of the ideas, objects, and technological knowledge along the littoral of the Uruguay River. The littoral zone would have been a "hot spot" area (Ord and Getis 1995; Trisalyn and Boots 2008) for the maintenance of social systems, which would include, among other things, conflicts, production, exchange, distribution and circulation of cultural materials. According to the techno-morphological characters present in the lithic materials recovered in the sites located at the coast of the Uruguay River and tributaries, dated as in the late Pleistocene-Holocene transition, the nodal role of the area can be traced back to more than 10,000 ^{14}C years BP (Austral 1995; Castiñeira 2008, 2009; Castiñeira et al. 2012; Dias Schmidt 2004; Dias Schmidt and Jacobus 2001; Hilbert 1991; Suárez 2004; Suárez et al. 2017).

Acknowledgements This study was funded by the Agencia Nacional de Promoción Científica y Tecnológica (ANPCyT) of Argentina (PICT 2014-0851 and PICT 2017-2723). We are very grateful for the help and information provided by Cristina Vassallo Cettour (Museum of Anthropological and

Natural Sciences of Concordia); for the logistical support provided by the town of Puerto Yeruá, especially Mr. Héctor Crisel and Mr. Guillermo Filsinger. We thank Mercedes Fernández, who generously helped us with the English translation of the manuscript.

References

Andrefsky W (1998) Lithics macroscopic approaches to analysis. Cambridge University Press, Cambridge

Antón D (1975) Evolución Geomorfológica del Norte del Uruguay. Ministerio de Agricultura Ganadería y Pesca, Montevideo

Apolinaire E (2016) Arqueología del suroeste de Entre Ríos: tecnología, subsistencia e interacción social en tiempos prehispánicos. Unpublished Doctoral Dissertation, Universidad Nacional de La Plata, La Plata, Argentina

Apolinaire E (2018) Characteristcs of secondary lithic deposits at the lower Uruguay River: applications for archaeological contexts in the inland plains of northeastern Argentina. Bol Mus Emílio Goeldi Cienc Hum 13(2):443–457

Apolinaire E, Bastourre ML (2016) Los documentos históricos de los primeros momentos de la Conquista del Río de La Plata (SXVI-XVII): una síntesis etnohistórica comparativa. Relaciones de la Sociedad Argentina de Antropología 41(2):319–351

Aschero C, Hocsman S (2004) Revisando cuestiones tipológicas en torno a la clasificación de artefactos bifaciales. In: Acosta A, Loponte D, Ramos M (eds) Temas de Arqueología. Universidad Nacional de Luján, Buenos Aires, Análisis lítico, pp 7–25

Austral A (1995) Los cazadores del sitio estratificado Pay Paso hace 10,000 años. In: Consens M, López JM, Curbelo C (eds) Arqueología en el Uruguay. MEC, Montevideo, pp 212–218

Arzamedia V, Giraudo A, Bellini G (2015) Biogeographical relationships of the large rivers of the Plata Basin based on snake assemblages. Revista Mexicana de Biodiversidad 86:674–684

Baeza J, Taddei A, Femenías J, Rodríguez O, Melgar W, Díaz A, Fornaro M (1977) Investigaciones Arqueológicas en el Área de Salto Grande: Tres Primeros Datos Radiocarbonos. V Encuentro de Arqueología del Litoral. Ministerio de Educación y Cultura and Intendencia Municipal de Río Negro, Río Negro, pp 67–88

Balfet H, Fauvet-Berthelot M, Monzón S (1992) Normas para la descripción de vasijas cerámicas. Centro de Estudios Mexicanos y Centroamericanos, México F.D

Berberian E, Nielsen A (2001) Historia Argentina Prehispánica. Brujas, Córdoba

Blasi A, Castiñeira C, Cusminsky G, Carignano A (2016) The Marine Isotopic Stage 3 (MIS 3) in Valleys of the Undulated Pampa, Buenos Aires Province, Argentina. In: Gasparini G, Rabassa J, Deschamps C, Tonni E (eds) Marine Isotope Stage 3 in Southern South America, 60 ka BP–30 ka BP, Springer Earth System Sciences. Springer, Dordretch, pp 129–146

Bombín M (1975) Afinidade paleoecológica, cronológica e estratigráfica de componente megamamífero na biota do Quaternário terminal da província de Buenos Aires (Argentina), Uruguai e Rio Grande do Sul (Brasil). Com do Museu de Ciências PUCRGS 3:1–28

Bombín M (1976) Modelo paleoecológico evolutivo para o neoquaternário da região da Campanha-Oeste do Rio Grande do Sul (Brasil): a Formação Touro Passo, seu conteúdo fossilífero e a pedogênesepós-deposicional. Com do Museu de Ciências PUCRGS 5:1–190

Bonomo M, Costa Angrizani R, Apolinaire E (2015) A model for the Guaraní expansion in the La Plata Basin and littoral zone of southern Brazil. Quat Int 356:54–73

Boretto R, Bernal R, Schmitz P, Basile Becker I (1973) Arqueología en el Departamento de Río Negro (R.O. del Uruguay). Esquema tentativo de una secuencia cronológica para sítios del Río Uruguay y Río Negro. In: I Congreso Nacional de Arqueología, Segundo Encuentro de Arqueología del Interior, Museo Municipal de Historia Natural de Río Negro, Río Negro, pp 1–16

Bórmida M (1964) Las industrias líticas precerámicas del Arroyo Catalán Chico y el Río Cuareim (Depto. de Artigas, R. O. Uruguay). Revista di Scienze Prehistoriche 19:195–232

Borrero L, Zárate M, Mitti L, Massone M (1998) The Pleistocene – Holocene Transition and Human Occupations in the southern cone of South America. Quat Int 48:191–198

Bracco R, Ures C, Ubilla M (2001) Contribución al análisis crítico de la relación espacial y temporal entre el material cultural y la paleofauna (Pleistoceno final y Holoceno) en la cuenca norte del Uruguay. Actas del IX Congreso Nacional de Arqueología Uruguaya. Asociación Uruguaya de Arqueología, Colonia, pp 327–340

Bueno L, Dias A (2015) Povoamento inicial da América do Sul: contribuições do contexto brasileiro. Estudos Avançados 29(83):119–147

Cabrera A (1953) Esquema fitogeográfico de la República Argentina. Revista del Museo de La Plata 8:87–168

Cabrera L (1994a) Quiénes habitaron el Uruguay en los últimos 10000 años? Aportes para el conocimiento de la prehistoria uruguaya. Ministerio de Educación y Cultura, Montevideo, pp 119–130

Cabrera L (1994b) Subsistema tecnológico y estrategias adaptativas en el Río Uruguay medio. In: Lanata J, Borrero L (eds) Arqueología de Cazadores – Recolectores. Límites, Casos y Aperturas, Arqueología Contemporánea 5:41–49

Capdepont I (2012) Arqueología de Sociedades indígenas del litoral del río Uruguay. Unpublished Doctoral Dissertation, Universidad. del Centro de la Provincia de. Buenos Aires, Olavarría

Capdepont I, Castiñeira C, Blasi A, del Puerto L (2017) Conectividad social durante el Holoceno tardío en el paisaje arqueológico del litoral oriental del bajo río Uruguay. Pesquisas Antropología 73:93–120

Carver R (1971) Procedures in sedimentology. Wiley-Interscience, New York

Castiñeira C (2008) Aspectos de la colonización humana prehistórica del noroeste del Uruguay. Unpublished Doctoral Dissertation, Universidad de Buenos Aires, Buenos Aires

Castiñeira C (2009) Aspectos de la colonización humana prehistórica del noroeste del Uruguay. Arqueología 15:241–245

Castiñeira C, Fernicola JC (2005) Tefrocronología y avances para una reconstrucción paleoclimática del Holoceno Medio en la región norte del Uruguay. Anales de Arqueología y Etnología 60:229–252

Castiñeira C, Apolinaire E, Blasi A, Bonomo M, Politis G, Bastourre L, Mari F (2017) Pre-Hispanic settlements in hydrometeorologically susceptible areas during the late Holocene: the upper delta of the Paraná River case. The Holocene 27:1801–1811

Castiñeira C, Blasi A, Politis G, Bonomo M, del Puerto L, Huarte R, Carbonari J, Mari F, García Rodríguez F (2013) The origin and construction of pre-Hispanic mounds in the upper delta of the Paraná River (Argentina). Archaeol Anthropol Sci 5:37–57

Castiñeira C, Charlín J, Cardillo M, Baeza J (2012) Exploring morphometric variations in fishtail projectile points from Uruguay, Pampa and Patagonia. Curr Res Pleistocene 1:57–61

Castiñeira C, Zárate M, Blasi A, Fernicola JC, del Puerto L, Inda H, Bracco R, García-Rodríguez F (2011) Aportes para una actualización de la correlación entre la Fm. Sopas del norte de Uruguay – Fm. Luján de Provincia de Buenos Aires: implicancias arqueológicas. In: Cocco G, Feulliet Terzaghi (eds) Arqueología de Cazadores Recolectores en la Cuenca del Plata, Centro de Estudios Hispanoamericanos, Santa Fe, pp 209–222

Castro JC (2017) Investigaciones arqueológicas en la cuenca media e inferior del Río Uruguay (Provincia de Entre Ríos). Unpublished Doctoral Dissertation, Universidad Nacional de La Plata, La Plata

Castro JC, Costa Angrizani R (2014) El sitio arqueológico Ensenada del Bellaco (Gualeguaychú, Entre Ríos) como indicador de la presencia guaraní en el río Uruguay Inferior. Revista del Museo de Antropología 7(2):235–242

Cerutti C (1992) Cambios Climáticos y Poblaciones Prehistóricas del Litoral Argentino. In: Iriondo M (ed) El Holoceno en la Argentina. CADINQUA, Paraná, pp 39–49

Cerutti C (2000) Ríos y Praderas: Los Pueblos Del Litoral. In: Tarragó M (ed) Nueva Historia Argentina. Los Pueblos Originarios y La Conquista, Sudamericana, Buenos Aires, pp 107–146
Cerutti C (2013) Homenaje a Jorge A y Amílcar Rodríguez: el poblamiento temprano en la provincia de Entre Ríos, Argentina. Cuadernos del Instituto Nacional de Antropología y Pensamiento Latinoamericano 4:34–47
Cerutti C, Gonzáles MI (2007) Modos de vida vinculados con ambientes acuáticos del nordeste y pampa bonaerense de Argentina. Relaciones de la Sociedad Argentina de Antropología 32:101–140
Cigliano E (1968) Investigaciones Arqueológicas en el Río Uruguay Medio y Costa Noreste de la Provincia de Buenos Aires. Pesquisas 18:5–9
Constantine P (2014) Exploring the spatial structure of pre-columbian cultural landscapes in the alto Paraná (Misiones Province, Argentina). Unpublished Doctoral Dissertation, University of Southampton, Southampton, UK
Costa Angrizani R (2012) Variabilidad, movilidad y paisaje. Una propuesta interpretativa para los vestigios de los asentamientos pre-coloniales en el noroeste del Río Grande do Sul (Brasil). Unpublished Doctoral Dissertation, Universidad Nacional de La Plata, La Plata
Cremonte M, Bugliani M (2009) Pasta, Forma e Iconografía. Estrategias para el estudio de la cerámica arqueológica. Xama 19(23):239–262
Day R (1965) Particle fractionation and particle size analysis. In: Black C (ed) Methods of Soil Analysis. American Society of Agronomy, Madison, pp 545–567
del Puerto L (2011) Silicofitolitos como indicadores paleoambientales. Bases comparativas y Reconstrucción paleoclimática a partir del Pleistoceno tardío en el SE del Uruguay. Editorial Académica Española, Berlín
del Puerto L, García-Rodríguez F, Bracco R, Castiñeira C, Blasi A, Inda H, Mazzeo N, Rodríguez A (2011) Evolución climática holocénica para el sudeste de Uruguay análisis multy proxy en testigos de lagunas costeras. In: García-Rodríguez F (ed) El Holoceno en la zona costera de Uruguay. Departamento de Publicaciones UdelaR, Montevideo, pp 117–147
del Puerto L, Gianotti C, Inda H (2016) Gestión del medio y producción de recursos en las Tierras Bajas del noreste de Uruguay: análisis paleoetnobotánico del sitio Pago Lindo. CAPA 13(25):198–222
Díaz A (1977) Arqueología de Salto Grande: secuencia cultural resultante de las investigaciones realizadas en Isla de Arriba e Isla del Medio (Uruguay). V Encuentro de Arqueología del Litoral. Ministerio de Educación y Cultura and Intendencia Municipal de Río Negro, Río Negro, pp 155–164
Dias Schmidt A, Hoeltz S (1997) Proposta metodológica para o estudo das indústrias líticas do sul do Brasil. Revista do CEPA 21(25):21–62
Dias Schmidt A (2004) Diversificar para poblar. El contexto arqueológico brasileño en la transición Pleistoceno-Holoceno. Complutum 15:249–263
Dias Schmidt A, Jacobus A (2001) The antiquity of the peopling of southern Brazil. Current Research in the Pleistocene 18:17–19
Favier Dubois C, Zárate M (2012). Breve historia geológica y climática. In: Otero H (ed) Población, Ambiente y Territorio, Unipe, La Plata, pp 51–78
Ferring R (2017) Alluvial Setting. In: Gilbert A, Goldberg P, Holliday V, Mandel R, Sternberg (eds) Encyclopedia of geoarchaeology. Springer, New York, pp 4–13
Fidalgo F, De Francesco F, Colado U (1973) Geología superficial de las hojas Castelli, J. M. Cobo y Monasterio (Provincia de Buenos Aires). V Congreso Geológico Argentino, Asociación Geológica Argentina, Buenos Aires 4:27–39
Flegenheimer N, Bayón C, Valente M, Baeza J, Femenías J (2003) Long distance tool stone transport in the Argentine Pampas. Quat Int 109–110:49–64
Flint R (1971) Glacial and quaternary geology. Willey Edition, New York
Folk R (1954) The distinction between grain size and mineral composition in sedimentary-rock nomenclature. J Geol 62(4):344–359

Gilbert A, Goldberg P, Holliday V, Mandel R, Sternberg R (eds) (2017) Encyclopedia of geoarchaeology. Springer, Heidelberg

Guidón N (1984) Misión de Rescate Arqueológico de Salto Grande. República Oriental del Uruguay, Ministerio de Educación y Cultura, Montevideo

Guidón N (1989) Misión de Rescate Arqueológico de Salto Grande, República Oriental del Uruguay, Tomo 2. Primera Parte, Ministerio de Educación y Cultura, Montevideo

González del Tánago M, García de Jalón D (1998) Restauración de Ríos y Riberas. Escuela Técnica Superior de ingenieros de Montes, Madrid

Gónzales-Ruibal A, Hernando A, Politis G (2011) Ontology of the self and material culture: arrow-making among the Awá hunter-gatherers (Brazil). J Anthropol Archaeol 30:1–16

Goso A, Ubilla M (2004) Los depósitos continentales cuaternarios del Norte de Uruguay: estratigrafía y paleontología. IV Congreso Uruguayo de Geología, Asociación Uruguaya de. Geología, Montevideo

Hally D (1983) Use alteration of pottery vessels surfaces: an important source of evidence for identification of vessel function. N Am Archaeol 4(1):2–26

Haynes C (1964) The earliest Americans. Science 166:709–715

Heusser CJ (1974) Vegetation and climate of the southern Chilean lake district during and since the last interglaciation. Quat Res 4:290–315

Hilbert K (1985) Archäologische Fundplätze des Rio Uruguay, Tigre und des Mendigú, Republik Uruguay. Beiträge zur Allgemeinen und Vergleichenden Archäologie 7:447–561

Hilbert K (1991) Aspectos de la Arqueología en el Uruguay. Mainz am Rhein: von Zabern (ed) Materialien zur Allgemeinen und Vergleichenden Archäologie, Germany

Hill C (2017) Geoarchaeology, History In: Gilbert A, Goldberg P, Holliday V, Mandel R, Sternberg (eds) Encyclopedia of geoarchaeology. Springer, New York, pp 292–303

Hogg A, Hua Q, Blackwell P, Niu M, Buck C, Guilderson T, Heaton T, Palmer J, Reimer P, Reimer R, Turney C, Zimmerman S (2013) SHCal13 southern hemisphere calibration, 0–50,000 years cal BP. Radiocarbon 55(4):1889–1903

Huarte R, Figini A (1988) La Plata radiocarbon laboratory liquids scintillation counting. Radiocarbon 30:347–350

Iriarte J, De Blasis P, De Souza J, Corteletti R (2017) Emergent complexity, changing landscapes, and spheres of interaction in southeastern South America during the middle and late Holocene. J Archaeol Res 25(3):251–313

Iriondo M (1996) Estratigrafía del Cuaternario de la Cuenca del Río Uruguay. In: Actas del XIII Congreso Geológico Argentino and III Congreso de Hidrocarburos. Asociación Geológica Argentina, Buenos Aires, vol IV, pp 15–25

Iriondo M, Kröling D (2004) The parent material as the dominant factor in Holocene pedogenesis in the Uruguay River Basin. Revista Mexicana de Ciencias Geológicas 21:175–184

Iriondo M, Kröling D (2008) Cambios ambientales en la cuenca del Uruguay. Universidad Nacional del Litoral, Colección Ciencia y Técnica, Santa Fe, Desde dos millones de años hasta el presente

Journaux A, Pellerin J (1980) Application des methods géomorphologiques aux recherché archéologiques: l'exemple des sites archéologiques de la region de Salto (Uruguay). Journal de la société des americanistes 67:215–225

Krieger A (1964) Early man in the new world. In: Jenning J, Norbek E (eds) Prehistoric man in the new world. Chicago Press, Chicago, pp 23–84

Lafón C (1971) Introducción a la arqueología del nordeste argentino. Relaciones de la Sociedad Argentina de Antropología 2:119–152

Laprida C, Orgeira MJ, García Chapori N (2009) El registro de la Pequeña edad de Hielo en las lagunas pampeanas. Revista de la Asociación Geológica Argentina 65(4):603–611

López JM (2013) Early human occupation of Uruguay: radiocarbon database and archaeological implications. Quat Int 301:94–103

Miller E (1976) Resultados preliminares das pesquisas arqueológicas paleoindígenas no Rio Grande do Sul, Brasil. In: Chiaramonte JC (ed) Actas del XLI Congreso Internacional de Americanistas, Instituto Nacional de Antropología e Historia de México, Mexico D F, vol 3, pp 484–491

Miller E (1987) Pesquisas arqueológicas paleoindígenas no Brasil ocidental. Estudios Atacameños 8:37–61

Morcote Ríos G, Mora Camargo S, Franky Calvo C (eds) (2006) Pueblos y paisajes antiguos de la selva amazónica. Universidad Nacional de Colombia, Bogotá

Nami H (1994) Paleoindio, cazadores recolectores y tecnología lítica en el extremo sur de Sudamérica continental. In: Lanata J, Borrero L (eds) Arqueología de cazadores recolectores. Limites, Casos y Aperturas. Arqueología Contemporánea 5:89–103

Nami H (2007) Research in the Middle Negro River Basin (Uruguay) and the Paleoindian Occupation of the Southern Cone. Curr Anthropol 48:164–174

Noelli F (2004) La distribución geográfica de las evidencias arqueológicas guaraní. Revista de Indias 64(230):17–34

Ord JK, Getis A (1995) Local spatial autocorrelation statistics: distributional issues and an application. Geogr Anal 27(4):286–306

Orton C, Tyers P, Vince A (1993) Pottery in archaeology. Cambridge University Press, Cambridge

Panario D, Gutiérrez O (1999) The Continental Uruguayan Cenozoic: an overview. Quat Int 62:75–84

Piovano E, Ariztegui D, Córdoba F, Cioccale M, Sylvestre F (2009) Hydrological variability in South America below the Tropic of Capricorn (Pampas and eastern Patagonia, Argentina) during the last 13.0 ka. In: Vimeux F, Sylvestre F, Khodri M (eds) Past climate variability from the last glacial maximum to the Holocene in South America and surrounding regions. Developments in paleoenvironmental research series. Springer, pp 323–351

Politis G (1991) Fishtail projectile points in the southern cone of South America: an overview. In: Bonnichsen R (ed) Clovis: origins and adaptation. Center for Study of the First American, University of Maine, Orono, pp 287–301

Politis G, Bonomo M (2012) La entidad arqueológica Goya-Malabrigo (ríos Paraná y Uruguay) y su filiación Arawak. Sociedade de Arqueología Brasileira 25(1):10–46

Politis G, Bonomo M, Castiñeira C, Blasi A (2011) Archaeology of the upper delta of the Paraná River (Argentina): mound construction and anthropic landscapes in the Los Tres Cerros locality. Quat Int 245(1):74–88

Politis G, Guitérrez M, Rafuse D, Blasi A (2016) The arrival of homo sapiens into the southern cone at 14,000 years ago. PLoS ONE 11(9):1–27

Politis G, Messineo P, Kaufmann C (2004) El poblamiento temprano de las llanuras pampeanas de Argentina y Uruguay. Complutum 15:207–224

Politis G, Prates L, Pérez S (2015) Early Asiatic migration to the Americas: a view from South America. In: Frachetti M, Spengler R (eds) Mobility and ancient society in Asia and the Americas. Springer, Cham, pp 89–102

Pouey V (2018) La ocupación cazadora – recolectora durante la transición Pleistoceno -Holoceno en el oeste de Rio Grande do Sul – Brasil; geoarqueología de los sitios en la formación sedimentaria Touro Passo. Archaeopress Publishing LTD, Oxford

Quattrocchio M, Borromei A, Deschamps C, Grill S, Zavala C (2008) Landscape evolution and climate changes in the late Pleistocene-Holocene, southern Pampa (Argentina): evidence from palynology, mammals and sedimentology. Quat Int 181:123–138

Rodríguez J (1992) Arqueología del sudeste de Sudamérica. In: Meggers Betty (ed) Prehistoria Sudamericana. Nuevas Perspectivas, Washington, pp 177–209

Rodríguez JA, Rodríguez A (1984) Nota preliminar sobre investigaciones arqueológicas en el NE de Entre Ríos y SE de Corrientes: el sitio Rancho Colorado. Revista del Museo de La Plata 64(8):385–403

Rodríguez B, Weissel M (2013) Exploraciones Arqueológicas. Recuperación del patrimonio hispánico guaraní en el corredor del río Uruguay. Consejo Federal de Inversiones de la Provincia de Entre Ríos, Buenos Aires, pp 63–80

Roper D, Josephs R, Beck M (2010) Determining provenance of shell-tempered pottery from the Central Plains using petrography and oxidation analysis. Am Antiq 75:134–157

Serrano A (1932) Exploraciones arqueológicas en el río Uruguay Medio. Memorias del Museo de Paraná 2:1–89

Serrano A (1950) Los primitivos habitantes de Entre Ríos. Ministerio de Educación y Cultura, Paraná

Serrano A (1972) Líneas fundamentales de la arqueología del litoral (una tentativa de periodización). Universidad Nacional de Córdoba, Córdoba

Shott M (2003) Chaîne opératoire and reduction sequence. Lithic Technology 28(2):95–104

Stern C (1990) Tephrochronology of Southernmost Patagonia. Natl Geogr Res 6:110–126

Stern C (1991) Mid-Holocene tephra on Tierra del Fuego (54°S) derived from Hudson Volcano (46°S): evidence large explosive. Revista Geológica de Chile 18(2):139–146

Stern C (1992) Tefrocronología de Magallanes: Nuevos datos e implicaciones. Anales del Instituto de Patagonia 21:130–141

Stuiver M, Polach H (1977) Discussion reporting of ^{14}C data. Radiocarbon 19(3):355–363

Suárez R (2000) Paleoindian occupations in Uruguay. Curr Res Pleistocene 17:78–80

Suárez R (2003) Paleoindian components in north Uruguay: new data on early human occupations of the late Pleistocene and early Holocene. In: Miotti L, Salemme M, Flegenheimer N (eds) Where the south winds blow. Center for the Study of the First Americans, Austín, pp 29–36

Suárez R (2004) La arqueología de los primeros americanos en Uruguay. Componentes paleoindios de los Ríos Uruguay – Cuareim y asociación entre cazadores humanos y fauna pleistocena en el sitio Pay Paso 1. In: Beovide L, Curbelo C, Barreto I (eds) La Arqueología ante los desafíos del nuevo siglo, Cd, Montevideo. ISBN: 9974–7811-08

Suárez R (2015) Adaptaciones Humanas tempranas y dispersión de la población durante la colonización del Río de la Plata y Río Uruguay. In: United Nations Educational, Scientific and Cultural Organization (ed) Human origin sites and the world heritage convention in the Americas, México DF, pp 161–177

Suárez R, Piñeiro G, Barcelo F (2017) Living on the river edge: the Tigre site (K-87) new data and implications for the initial colonization of the Uruguay River basin. Quat Int 473:242–260

Taddei A (1964) Un yacimiento precerámico en el Uruguay. Baessler-Archiv 12:317–372

Taddei A (1980) Industrias líticas del Uruguay y su relación con Pampa-Patagonia de Argentina. Revista de Arqueología 3(19):58–64

Tonello M, Prieto A (2010) Tendencias climáticas para los pastizales pampeanos durante el Pleistoceno tardío – Holoceno: estimaciones cuantitativas basadas en secuencias polínicas fósiles. Ameghiniana 47(4):478–501

Trisalyn A, Boots B (2008) Detecting spatial hot spots in landscape ecology. Ecography 31:56–566

Ubilla M (1996) Paleozoología del cuaternario continental de la cuenca norte del Uruguay: Biogeografía, cronología y aspectos climático – ambientales. Unpublished Doctoral Dissertation, Universidad de la República, Montevideo

Ubilla M, Perea D (1999) Quaternary vertebrates of Uruguay: biostratigraphic, biogeographic and climatic overview. Quat S Am Antarct Penins 12:75–90

Ubilla M, Corona A, Rinderknecht A, Perea D, Verde M (2016) Marine isotope stage 3 (MIS 3) and continental beds from Northern Uruguay (Sopas Formation): paleontology, chronology, and climate. In: Gasparini G, Rabassa J, Deschamps C, Tonni E (eds) Marine isotope stage 3 in Southern South America, 60 ka BP–30 ka BP. Springer Earth System Sciences, pp 183–205

Ubilla M, Perea D, Goso C, Lorenzo N (2004) Late Pleistocene vertebrates from northern Uruguay: tool for biostratigraphic climatic and environmental reconstruction. Quat Int 114:129–142

Zucol A, Colobig M (2014) Análisis de fitolitos. Propuesta de tratamiento metodológico y taxonómico. In: Avances y desafíos metodológicos en Arqueobotánica: Miradas consensuadas y diálogos compartidos desde Sudamérica, Universidad Internacional, Santiago, pp 232–251

Archaeomagnetism Applied to Ceramics from Coastal Archaeological Sites in Uruguay

Irina Capdepont, Leda Sánchez Bettucci, Juan Morales and Avto Gogichaishvili

Abstract The archaeological record of pottery in Uruguay has a wide spatial and chronological distribution. It extends along coastal areas from the west (Uruguay River) to southeast (Atlantic Ocean), and stretches from the mid-Holocene period until the arrival of the Europeans. The chronologies mark a continuity in pottery production from Prehistory (~3000 ^{14}C yr BP) to historical times. The wide spatial and temporal distribution of ceramics has promoted the debate on its origin and dispersion, still under discussion. In this instance, it was the intention to contribute, from the first application of archaeomagnetism on ceramic materials of prehistoric contexts in Uruguay, with direct data on ceramics chronology. Likewise, it is aimed at evaluating the reliability of the method by comparing it with other numerical age estimation techniques (^{14}C and TL). The results achieved for the fragments from La Esmeralda, Cráneo Marcado, and Guayacas sites provided highly reliable dating for pottery, between ~1400 years BC and ~1500 years BC. These ages are consistent with ^{14}C and TL datings. Likewise, the results show that archaeomagnetism is a useful technique in obtaining ceramics chronologies and contributes to the chronological discussion of its origin and dispersion in the coastal territory of Uruguay.

Keywords Pottery chronology · Archaeomagnetic dating · Holocene · Uruguay

---

I. Capdepont (✉)
Laboratorio de Estudios del Cuaternario (MEC-UNCIEP, IECA Facultad de Ciencias), Centro Universitario Regional del Este (CURE-Rocha), Universidad de la República, Iguá 4225, 11400 Montevideo, Uruguay
e-mail: i.capdepont@gmail.com

L. Sánchez Bettucci
Laboratorio de Geofísica y Geotectónica, Facultad de Ciencias, Departamento de Geodinámica Interna, Universidad de la República, Iguá 4225, 11400 Montevideo, Uruguay

J. Morales · A. Gogichaishvili
Laboratorio Interinstitucional de Magnetismo Natural, Instituto de Geofísica, Universidad Nacional Autónoma de México (UNAM), Unidad Michoacán, Campus Morelia, Morelia, Mexico

H. Inda Ferrero and F. García Rodríguez (eds.), *Advances in Coastal Geoarchaeology in Latin America*, The Latin American Studies Book Series,
https://doi.org/10.1007/978-3-030-17828-4_9

Introduction

The present work is part of the investigation on the pre-Hispanic ceramic societies that occupied the current coastal territory of Uruguay. The aim of this work is to study and obtain chronological data, thus contributing to the temporary location of pottery group settlements on the eastern coastal region of the Uruguay River (western region), the Atlantic coast and the lagoon coastal region (eastern region). The adoption of pottery resulted in variations in the technological system, as well as in the subsistence of the indigenous groups that manufactured it. It is aimed to fine-tune its temporal contextualization. From this perspective, in-depth studies are made on the spatiotemporal context of ceramics through the application, on different fragments, of correlation methods, specifically the archaeomagnetic technique. Archaeomagnetic dating offers an important chronological method for archaeological research, which is used in different parts of the world (e.g., Aidona and Kondopoulou 2012; Batt et al. 2017; Lengyel 2017; Novillo Verdugo et al. 2016, among others).

In order to establish the age of the cultural materials, as well as their origin, various physical techniques have been applied to ceramics over time. These currently use the variations of the Earth's magnetic field recorded in archaeological materials. Archaeomagnetic dating, resulting from the use of magnetic field variation, began to be applied in the 1930s in France, and in 1960, it became a reliable and interdisciplinary dating technique (Batt et al. 2017; Hus et al. 2003; Outram and Batt 2010). However, in Uruguay, this technique has only been recently implemented. This dating technique is based on the comparison of magnetic parameters of an archaeological material against a known record of geomagnetic variation. The application of non-conventional dating methods, according to Genevey and Gallet (2002), has led in recent years to a rapid development of archaeomagnetic methods.

Two key phenomena are considered when obtaining archaeomagnetic data: one is the Earth's magnetic field (which varies in space and time) and the other is that there are certain components in archaeological materials (ferromagnetic minerals in clays) that are capable of recording the values of variation in intensity and direction of the geomagnetic field experienced in the past (Lengyel 2017; Parés 2017). In this regard, some iron oxides (ferromagnetic minerals), which exhibit exceptional properties when cooled after having been exposed to a high-temperature heating process (500–700 °C), can be oriented in the direction of the geomagnetic field existing during cooling and steadily preserve over time the orientation acquired. The ability of archaeological objects to faithfully record such directional variations and intensity of the geomagnetic field in the past was disclosed in the late 1950s (Thellier and Thellier 1959).

Chronologies obtained with the archaeomagnetic dating method, when applied to materials from different archaeological sites in America (e.g., Hartmann et al. 2007; Morales et al. 2012; Novillo Verdugo et al. 2016) and Europe (e.g., Carmona et al. 2014; Ech-Chakrouni et al. 2013; Langouet and Goulpeau 1984; Prevosti et al. 2013), have provided reliable data on the temporal context of manufacture and use of pottery.

In this regard, an interdisciplinary verification of their suitability and chronological resolution in ceramic artifacts from archaeological sites in Uruguay has begun.

Temporality of the Uruguayan Pottery Record

The age and temporal context of the ceramic record has been one of the main research areas in the history of archaeology (e.g., Kolb 2017; Orton et al. 1997; Shepard 1954). Before accessing radiocarbon techniques, the pottery record was used to assign relative ages in the chronological reconstructions of a locality, an area, and even a region. Pottery, used as a diagnostic element, led to establishing settlement models in Uruguay with estimated time limits (e.g., Austral 1977; Díaz and Fornaro 1977; Hilbert 1991; Schmitz et al. 1991; Serrano 1972). The temporary reconstructions began to be more accurate with the access to radiocarbon dating techniques, which in Uruguay became mainstream in the 1990s (Bracco 1991; Bracco and Ures 1999). With this method, the temporal context of archaeological sites and indirectly, of the materials found therein, is specified (e.g., Beovide and Malán 2009; Capdepont 2001; Castillo 2001; Farías et al. 2001; López and Castiñeira 2001).

For the western region, different cultural models and representations were made based on ceramic materials, the Serrano (1972) model being the most widely disseminated in the region. This author elaborated a cultural development system based on the characterization and classification of ceramic types. This helped generate chronological sequences where the stylistic attributes and the variability of the ceramics are the result of different ethnic groups. In the archaeological bibliography over the past 30 years, it is observed how the cultural scheme proposed by Serrano (1972), albeit questioned and updated, is taken up by several researchers (e.g., Capdepont 2013; Castillo 2001; Díaz and Fornaro 1977; Guidón 1989; Politis and Bonomo 2012). In general, this model maintains its validity and includes the following cultural entities: Entrerriana or Básica del Litoral (formed by fishing and hunting groups, ca. 2500 ^{14}C years BP), Ribereños Plásticos (attributed to the Chaná-Timbú groups ca. 1500 ^{14}C years BP), and Tupiguaraní (ca. 500 ^{14}C years BP) (Capdepont 2013). However, the presence of pottery as a chronological reference guide to assign sites of the western region a temporality no older than the one stated above (2500 ^{14}C yr BP) has been questioned in recent years. According to new data, the beginning of pottery in the region dates back to more than 4000 ^{14}C yr BP. The ceramics from the La Tuna site in the San José Department, according to Beovide and Malán (2009) and Beovide and Campos (2014), were recovered from a context dated to 4800 ^{14}C yr BP. Likewise, the ceramic context of the Yacaré Cururu site, Artigas Department, is linked to a chronology of 3800 ^{14}C yr BP (Consens 2007). However, later ceramic records are better represented at a regional level, while these types of records for mid-Holocene are still scarce.

The ceramic material from the eastern region has been attributed to the so-called Vieira tradition (sensu Steward 1946). The attribution to this archaeological tradition relates to explanatory models based on time sequences (sorted in traditions and phases). These sequences have been elaborated with ceramic material using the Ford method (Schmitz 1991; Schmitz et al. 1991), and are directly related to the archaeological record of the Indian mounds (anthropic constructions on land) and have been initially characterized as simple, homogeneous, and utilitarian in nature (e.g., Bracco et al. 2000a; Femenías 1990; Naue 1973; Schmitz 1968; Schmitz et al. 1991). According to new interpretive perspectives and research advances in eastern Uruguay, a higher level of complexity for Indian mounds and the cultural materials found in them has been suggested (e.g., Bracco et al. 2000b; López 2000; Pintos and Capdepont 2001). The stratigraphic levels of the Indian mounds (Cerritos) with the presence of pottery are chronologically between ca. 1500 and 3000 ^{14}C yr BP (Bracco et al. 2000a; Capdepont 2001; Capdepont et al. 2016; Hilbert 1991; López 2001; López and Castiñeira 2001; López et al. 2014; Schmitz et al. 1991).

Studies carried out on the ceramic material, from a broad and inclusive perspective, have allowed to explore the different aspects related to its manufacture, use and disposal, as well as to distinguish the links and social connectivity that could have taken place between the pottery populations during the Holocene (Capdepont et al. 2017). Most of the pottery sites in the Uruguayan territory have been located in late Holocene contexts (Baeza et al. 1977; Capdepont 2001, 2013; Castillo 2001, 2004; Díaz and Fornaro 1977; Díaz 1977; Farías et al. 2001; López 2001; López and Castiñeira 2001, among others). These contexts coincide, among other things, with the management of some cultivated plants, the presence of polished artifacts, intense human burials and *Cannis familiaris* funerary activity, as well as an improvement in climatic conditions. This suggests a temperate and humid environmental condition in the late Holocene, which according to del Puerto et al. (2012) contributed to technological and subsistence changes, as well as further intensification in the occupation of spaces that were not available until ca. 3000 years BP (Capdepont 2013).

Techno-morphological, stylistic, and functional ceramic variability recorded in both regions, as well as the broad time span proposed for pottery sites from ca. 4500 years BP to the sixteenth century, encourages further study of direct dating of the ceramic material. The aim is to adjust origin and dispersion chronologies of ceramic in the territory, as well as to discuss its contexts of association. Similarly, the chronological data can contribute to the discussion on the distribution over wide territorial scales and the temporality of the technological and subsistence changes that have taken place (starting with the inclusion of pottery) in the sociocultural development of the peoples who inhabited the current Uruguayan territory.

Archaeological and Pottery Sites

Archaeological sites, from which the materials have been recovered, are located in two regions of the Uruguayan territory: east and west. The sites of the eastern region are located in the Rocha Department and are linked to developed settlements near the Atlantic coast. This coastal area will be represented by evidence from the La Esmeralda and Cráneo Marcado sites. On the other hand, the evidence that comes

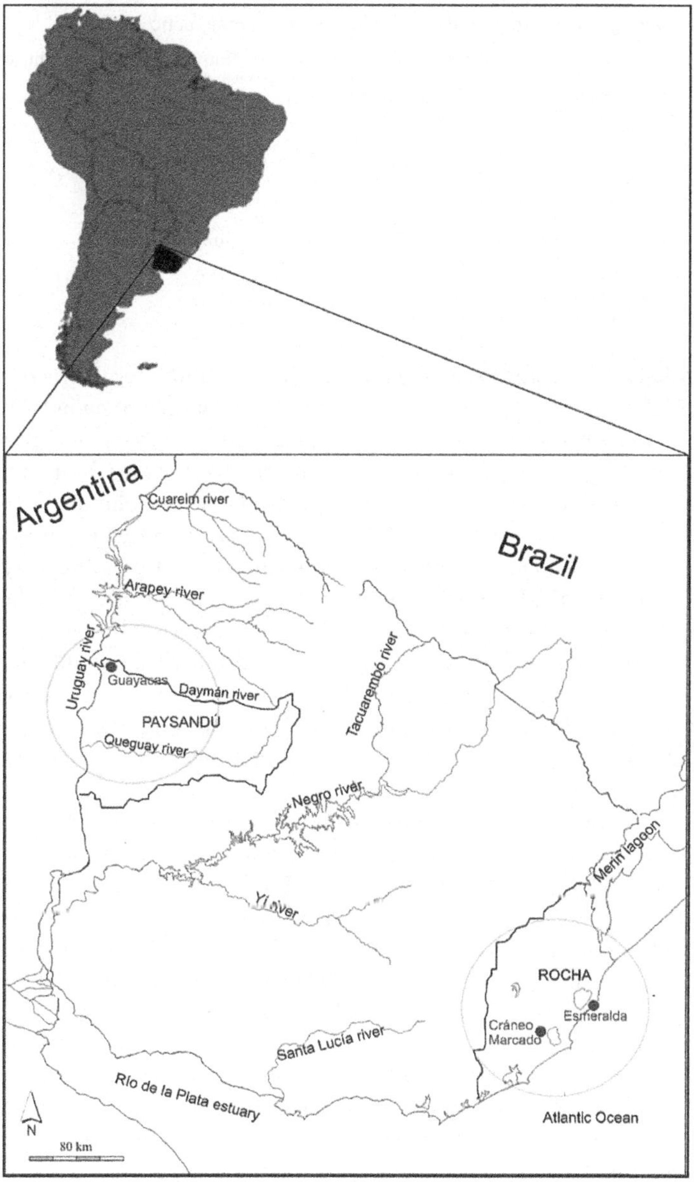

Fig. 1 General location of archaeological sites of the western (left) and eastern (right) regions of Uruguay

from the Guayacas site, located in the Paysandú Department, represent the pottery record of the western region, linked to the development of settlements in the fluvial littoral (Fig. 1 and Table 1).

Table 1 Samples of pottery fragments processed for archaeomagnetic analysis

Sample	Region/Department	Site location	Name	Context
E05	East-Rocha	Atlantic coast	La Esmeralda	Surface collection
E11				
CM01		Lagoon coast	Cráneo Marcado	Excavation
CM03				
Gy06033	West-Paysandú	Riverbanks	Guayacas	
Gy7				

Archaeological research conducted in both regions allow for the recovery of different cultural materials (e.g., pottery, lithic, plant, and animal remains). These materials provide evidence of human activities developed by indigenous groups during the Holocene (Table 2). It is on the pottery material, recovered in the above sites, that the studies were performed using the archaeomagnetic technique.

La Esmeralda site, located on the Atlantic coast, is composed of three shell middens (an accumulation of anthropogenic biogenic material, mainly *Donax hanleyanus* Philippi 1847 is described) with the presence of cultural materials. They show human settlement and exploitation of marine resources since ca. 3200 ^{14}C yr BP (Castiñeira et al. 2010; López et al. 2009). Associated with the group of shell middens are interdune spaces with significant dispersion of cultural materials (Fig. 2). The ceramic sample analyzed in the present work was recovered through the prospecting and surface collection activities carried out in the interdune spaces (sample E05 and E11).

The Cráneo Marcado site, located on the western banks of Castillos Lagoon, has two different occupation areas. On the one hand, at a height of 20 m a.s.l., on late Pleistocene–Holocene glacis, three mound structures (anthropic constructions on land) are located and, at a distance of 1000 m, a parallel plain extends near the current coastline (Capdepont et al. 2016) (Fig. 3). The concentration of archaeological materials in the plain led to excavation I, where the ceramic material studied was recovered (sample CM01). These materials are associated with developed soils on storm bioclastic crests, located from the vicinity of 3.5 m a.s.l. The material is recovered from a context dominated by fauna post-consumption discarded remains and dated to ca. 3000 ^{14}C yr BP (Pintos and Capdepont 2001). One of the ceramic fragments studied in this work (CM03) was recovered from one of the circular mound structures (mound B), 35 m in diameter and 1.30 m in height. These are part of a set of cultural materials recovered in one of the layers of anthropic accumulation of ca. 2700 ^{14}C yr BP (Capdepont et al. 2016).

Research in the western region, on the banks of the Uruguay River, made it possible to locate several ceramic sites (Capdepont 2013). Between these sites is located the Guayacas site, on top of a hill on by the Daymán River, 10 km east of Uruguay River (Fig. 4). Lithic and ceramic/pottery material was recovered from the archaeological excavations. Their analysis shows a domestic context activity in a residential camp

Table 2 Chronological context of the analyzed material (chronologies (a) and (b) taken from Capdepont 2013; Capdepont et al. 2017; Castiñeira et al. 2010), data calibrated with Calib 7.0 (Hogg et al. 2013)

Site	Dated material	Chronologies obtained by (a) ^{14}C and (b) Thermoluminescence	Ages ^{14}C calibrated 1σ	Ages ^{14}C calibrated 2σ
La Esmeralda A	Charcoal	3190 ± 50 BP (URU0220) (a) 3060 ± 90 BP (URU0233) (a) 2360 ± 70 BP (URU0254) (a)	[BC1497: BC1318] [BC1393: BC1127] [BC510: BC232]	[BC1527: BC1269] [BC1494: BC1000] [BC749: BC201]
	Shellfish valves	3210 ± 50 BP URU0219) (a) 3010 ± 50 BP URU0234) (a) 2510 ± 50 BP (URU0253) (a)	[BC1503: BC1328] [BC1270: BC1090] [BC756: BC511]	[BC1603: BC1285] [BC1382: BC1014] [BC768: BC415]
La Esmeralda B	Shellfish valves	3190 ± 60 BP (URU0231) (a)	[BC1499: BC1308]	[BC1603: BC1232]
La Esmeralda C	Shellfish valves	1080 ± 60 BP (URU0230) (a)	[AC900: AC1130]	[AC890: AC1149]
	Charcoal	1000 ± 70 BP (URU0237) (a)	[AC1024: AC1154]	[AC972: AC1224]
Cráneo Marcado Plain	Charcoal	3050 ± 50 BP (URU136) (a)	[BC1376: BC1130]	[BC1408: BC1058]
Cráneo Marcado Mounds B	Charcoal	2760 ± 60 BP (GrA15608) (a)	[BC915: BC810]	[BC1006: BC795]
Guayacas	Pottery	1530 ± 11 AC (TL-Gy1) (b) 1471 ± 12 AC (TL-Gy2) (b) 1556 ± 18 AC (TL-Gy5) (b)	–	–

with a thermoluminescence age of ca. 1500 yr BC (Capdepont et al. 2013). The ceramic fragments analyzed in the present study (samples Gy06033 and Gy07) are part of two utilitarian vessels which, according to various analyses, are related to the processing of wild and cultivated plant-based foods (Capdepont 2013).

The pottery studied is considered as a set of forms produced by social action, so its recording has been addressed and understood in relationship with their context. Studies of ceramics have increased the knowledge of the operational technical chain, supply modes and manufacturing exploitation of resources used, as well as their use and recycling (Capdepont 2013; Capdepont and Castillo 2001). This analysis

Fig. 2 La Esmeralda archaeological site location: **a** interdune site area structures; **b** detail of the interdune space where the material studied was recovered; **c** fragment of the vessel from which pots E05 and E11 come

has been carried out through technological, morphological, functional, stylistic, and chronological studies. Also, they have allowed to deeper recognize into the vessels' life history (from the selection of raw materials for its production taking into account the manufacturing, use, discarding, and/or reuse processes).

Over time, pottery has been very important in the study of cultural habits related to lifestyles and economic and symbiotic systems. These material products, as well as the people who make them, are the result of the society in which they live. Given that the social products that are created in a group are related to all areas of their reality, it is understood that the characteristics and elements of a society are reflected in their material production (Cobas and Prieto 1998). Ceramics, as one of the material and

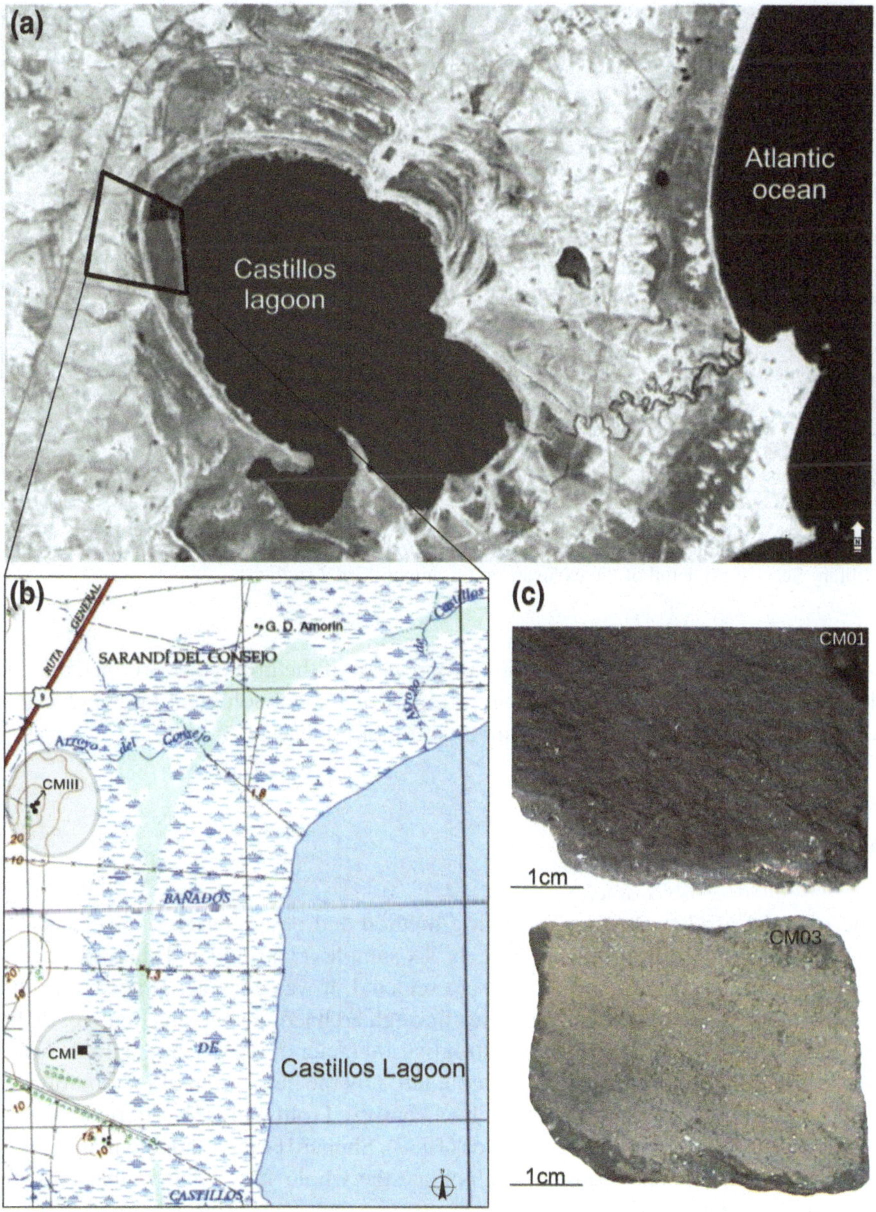

Fig. 3 Location of Cráneo Marcado site: **a** aerial photography (1:60,000 Military Geographical Service) with delimitation of the studied area; **b** detail of CMI site (plain) and CMIII (hill) on topographic map Sheet C26—Aguas Dulces (1:50,000 Military Geographic Service); **c** studied fragments CM01 and CM03

Fig. 4 Guayacas site location: **a** study area delimitation in aerial photography (1:20,000 Geographic Military Service); **b** detail of the excavation; **c** fragment Gy07 studied

social products, responds to the rationality pattern of the group that produces it, and the resources used in its manufacture are part of its technological knowledge that allows to place it in time through direct dating of pottery.

Methodology

The material used for testing the dating method and obtaining chronologies came from dated contexts. From these contexts, six samples classified and characterized at macroscopic and microscopic level were selected. It was considered necessary that the samples used to obtain chronologies through archaeomagnetism (Lengyel 2017) were previously studied due to the destructive character of the technique. In this way, the destruction of the material can be countered by the generation of information. In view of this, the selected material has been analyzed following the criteria proposed by Kolb (2017), Orton et al. (1997), Rice (1987), Shepard (1954) and Shennan (1992), among others. This has required to elucidate the whole story of the material, from the initial production stages to the final process that creates the vessel, as well its deposition and subsequent alteration (see Capdepont 2013; Capdepont et al. 2016).

Technological, morphological, functional, and stylistic attributes of the material are generally determined by the role they had, and their analysis can shed light, for example, on the activities carried out on a given site (Henrickson and MacDonald 1983). The data needed for this was accessed by implementing different methodologies and analytical techniques (e.g., morphometric, petrographic, and XRD).

After the technological, morphometric, functional, and stylistic study, the processing of each sample (n = 6) consisted of sectioning in ~1 × 1 cm fragments. In this way, between six and seven sub-specimens were obtained, which were then pressed into cylindrical salt tablets, with similar dimensions to those of standard paleomagnetic samples. The magnetic properties of the material were determined by the use of a thermomagnetic balance (AVFTB) using the irregular and smaller fragments obtained during sectioning. For this, they were pulverized using an agate mortar, thus obtaining ~300 mg. The isothermal remnant magnetization (IRM) acquisition curves, reverse field, magnetic hysteresis, and thermomagnetic curve were then generated. Their analysis was carried out using the RockMagAnalyzer 1.0 software (Leonhardt 2006). The archaeointensity determination was carried out by using the Thellier–Coe-type method (Thellier and Thellier 1959; Coe 1967) in an ASC Scientific TD48-SC oven. Ten double-heating steps were performed, distributed along the whole temperature range used (25–540 °C), with a temperature reproducibility of ±2 °C between two heatings at the same temperature. In addition, the laboratory field strength was established at 45 ± 0.5 μT, and tests known as "pTRM checks" and "pTRM Tail checks" (Riisager and Riisager 2001) were added to the laboratory procedure. The analysis was carried out using the ThellierTool 4.11 software (Leonhardt et al. 2004). The magnetization intensity dependence with the cooling rate was investigated following a modification to the procedure of Chauvin et al. (2000), with a slow cooling time of 6 h between 540 and 20 °C.

For the TRM anisotropy corrections (ATRM), the proposals by McCabe et al. (1985), Selkin et al. (2000) and Chauvin et al. (2000), among others, were considered. In the present work, different specimens of the same sample were placed in six different positions at the time of being pressed (into salt tablets) in order to reduce the bias related to the ATRM effects. This way of reducing the bias has been experimentally verified (Morales et al. 2011). The archaeomagnetic dating was performed upon obtaining the corresponding archaeointensities, and the results were analyzed using the archaeo_dating archaeomagnetic dating tool, implemented in MATLAB by Pavón-Carrasco et al. (2011).

Results

The ceramic characterization of the east and west regions shows a differential distribution, mainly at a technological and stylistic level (Fig. 5). On the one hand, the pottery of the east region is generally characterized by heterogeneous compact pastes, with tempers of coarse to fine quartz sands, vessels made with the coiling method, usually fired in a reducing environment and incomplete oxidation, with smooth surfaces and simple forms (e.g., Bracco et al. 2000a; Capdepont 2001; Capdepont et al. 2016; Durán 1990; Hilbert 1991). In contrast, the pottery of the western region shows greater technological, morphological, and stylistic variability. This pottery is characterized by friable, laminar homogeneous pastes, with temper of chamotte (sherd temper), medium and fine quartz sands, spicules of sponge and iron oxide nodules,

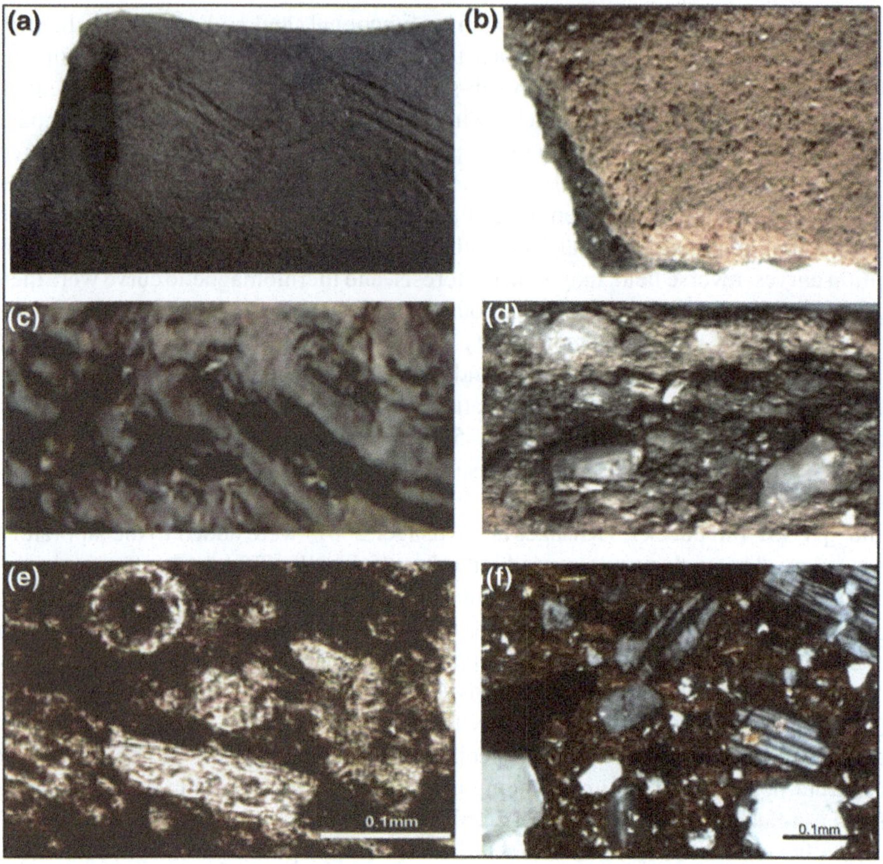

Fig. 5 Example of stylistic and technological differences between ceramics from the eastern (right) and western (left) regions: **a** incised decoration; **b** smooth surface; **c** type of homogeneous paste; **d** type of heterogeneous pulp; **e** fine-grained paste with sponge spicules observed in thin-film cut; **f** thin section petrography characterized by medium and coarse-grained paste with quartz and feldspar minerals

manufactured with the coiling and modeling method, fired in a reductive and oxidant environment, with various techniques and decorative styles in vessels of simple and complex shapes (e.g., Capdepont 2013; Capdepont and Castillo 2001; Castillo 2001; Díaz and Fornaro 1977; Durán 1990; Farías et al. 2001; Serrano 1972).

Morphological and bio-compositional analyses allowed to further analyze the functional variability of the vessels of both regions in connection with treatment of plant foods (see Capdepont 2013; Capdepont et al. 2016). The technical results, styles, forms, and decorative motifs in the western region are in contrast with the low variety observed in the pottery of the eastern region.

The archaeointensity results obtained in ceramic fragments, corrected by cooling rate, are shown in Table 3 and Fig. 6. These results suggest that samples from the eastern region (E05, E11, CM01, and CM03) have similar intensities. These intensities allowed to estimate a ca. 1200–1600 BC chronology (Table 4). The western region samples (Gy 06033 and Gy07) have similar intensities (Table 3) and suggest a chronology of ca. 1400–1600 AC, later than those of the east region (Table 4). These chronological estimates, presented in Table 4, would be supporting the origin of the materials of this archaeological context, as well as in the general regional context.

Table 3 Archaeointensity results

Sample	Tmin-Tmax (°C)	N	*f*	*g*	*q*	AI (μT)	1σ (μT)
E05	150–515	10	0.736	0.836	10.22	38.12	2.3
	150–515	10	0.818	0.864	15.36	36.18	1.67
	150–450	7	0.579	0.735	6.6	53.01	3.42
	150–515	9	0.696	0.81	9.35	33.57	2.03
	150–500	9	0.736	0.826	13.28	36.36	1.67
	150–515	11	0.744	0.875	8.14	31.5	2.52
					Aver =	**38.1**	**7.7**
E11	20–500	8	0.67	0.84	9.7	36.03	2.08
	350–515	6	0.44	0.78	7	38.04	1.84
	20–515	9	0.69	0.85	14.1	44.65	1.87
	300–500	6	0.44	0.78	9.4	37.81	1.38
	350–515	6	0.45	0.77	6	34.26	2
					Aver =	**38.2**	**3.9**
CM01	200–540	9	0.855	0.844	34.93	43.56	0.9
	150–540	10	0.88	0.861	19.58	45.36	1.76
	150–540	11	0.857	0.865	27.43	44.96	1.22
	150–500	9	0.78	0.808	22.67	43.7	1.22
	150–540	10	0.912	0.856	34.82	40.14	0.9
	150–500	8	0.743	0.763	20.09	39.87	1.13
					Aver =	**42.9**	**2.4**
CM03	150–540	12	0.984	0.858	16.89	40.5	2.03
	150–475	10	0.934	0.835	18.46	47.93	2.03
	150–540	10	0.972	0.822	22.46	35.42	1.26
	150–500	10	0.887	0.803	13.03	46.94	2.57
					Aver =	**42.7**	**5.9**
Gy06033	150–515	10	0.516	0.862	8.42	42.57	2.25
	150–540	12	0.652	0.849	7.75	41.58	2.97
	150–540	11	0.662	0.84	12.11	41.18	1.89
	150–500	9	0.524	0.847	5.45	51.39	4.19
	150–500	9	0.482	0.846	4.8	41.81	3.56

(continued)

Table 3 (continued)

Sample	Tmin-Tmax (°C)	N	*f*	*g*	*q*	AI (μT)	1σ (μT)
	200–515	9	0.467	0.852	9.16	37.31	1.62
					Aver =	**42.6**	**4.7**
Gy07	20–400	5	0.55	0.57	4.9	38.16	2.45
	20–400	5	0.62	0.56	3.2	46.34	5.04
	20–500	8	0.73	0.77	4	35.72	4.97
		0	–	–	–	N/R	–
		0	–	–	–	N/R	–
		0	–	–	–	N/R	–
		0	–	–	–	N/R	
					Aver =	**40.1**	**5.6**

References Sample: pot identifier; Tmin/Tmax: minimum/maximum temperature used for the adjustment by least squares; N: number of points used for the adjustment; f, g and q: quality parameters, defined by Coe et al. (1978); AI: archaeointensity; σ: standard deviation

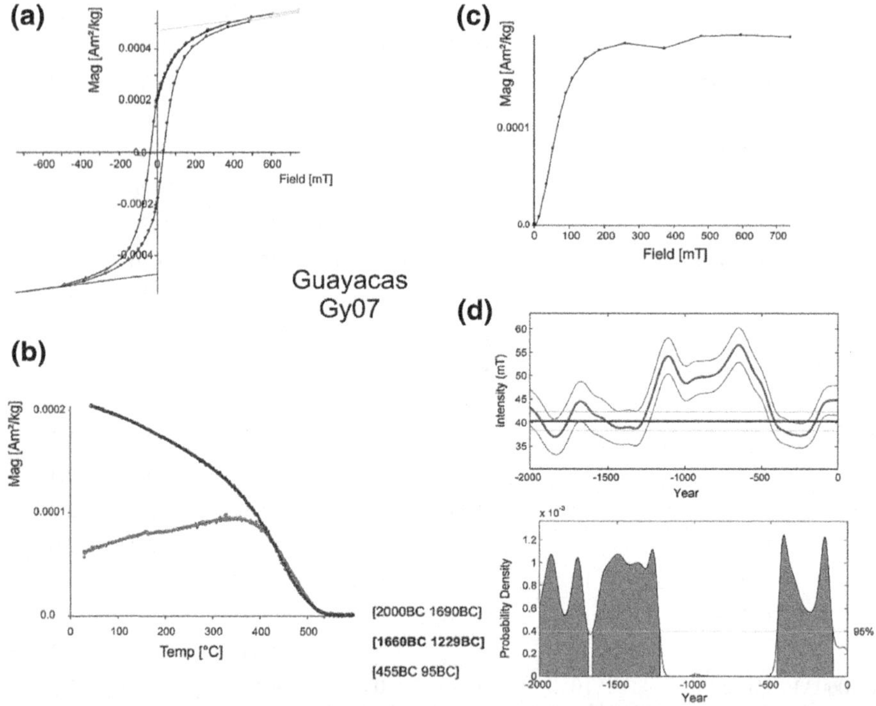

Fig. 6 Representative example of the results obtained for sample Gy07 (pots 07 of Guayacas): **a** magnetic hysteresis curve (pot-bellied type), **b** IRM acquisition curve, **c** thermomagnetic curve (M-T), and **d** archaeomagnetic dating

Table 4 Results with the most probable age ranges for each sample

Sample	Site	Average intensity (μT)	Age
E05	Esmeralda	38.1 ± 7.7	1257 BC–1584 BC
E11		38.2 ± 3.9	1286 BC–1469 BC
CM01	Cráneo Marcado Planicie	42.9 ± 2.4	1238 BC–1629 BC
CM03	Cráneo Marcado Mound B	42.7 ± 5.9	1226 BC–1660 BC
Gy06033	Guayacas	42.6 ± 4.7	1497 AC–1636 AC
Gy07		40.1 ± 5.6	1511 AC–1627 AC

Conclusions

The ceramic materials analyzed belong to vessels that have been used in day to day, domestic contexts (processing, cooking, storage, and consumption of food) during the Holocene. The ceramics of both coastal regions present distinctive characteristics related to the manufacturing process, the materials used and the place where they were made and used. These differences, reflected in the pastes, shape, technology, and decoration of the sherds, lead to believe that the groups producing the vessels in the east and the west did not share the same technological knowledge. Each group chose how to make the manufacture of a container and that choice has left different traces in the system of representations through time and space.

The data obtained allow to adjust the chronologies of the indigenous pottery, providing a reliable time frame for the materials of eastern and western Uruguay. Based on the archaeointensity results and regional archaeological contexts, the most probable chronology can be estimated between ~1400 years BC and ~1500 years AC for the ceramics studied from the east and west regions, respectively. Temporal differences are in line with those analyzed at a technological and stylistic level in both regions (e.g., Capdepont and Castillo 2001; Durán1990; Hilbert 1991). Also, the data suggest that the pottery groups operated in the territory during the late Holocene, a time also associated with an occupational increase and changes in subsistence in connection with climate changes (e.g., Capdepont 2013; del Puerto et al. 2016; López 2001).

The ages obtained closely correlate with the ^{14}C ages reported for the La Esmeralda and Cráneo Marcado sites, as well as those acquired by thermoluminescence (TL) for the Guayacas sherds (Tables 2 and 4). Consequently, the first conclusion is that the archaeomagnetic method is suitable for working with the ceramic temporality, as suggested by different authors, such as Batt et al. (2017), Lengyel (2017) and Parés (2017). Its application allows to obtain data that contribute to the chronological location of the pottery, as well as to have a better the temporal contextualization of the activities developed in the sites, such as technological and subsistence changes, among others. Archaeomagnetic dating is a valuable tool for determining chronologies in archaeological studies.

The data obtained show the potential of the method in solving archaeological problems. This stimulates the use and expansion of the archaeomagnetic studies of the ceramic record of various archaeological sites in Uruguay. The application of this technique can enhance the knowledge of the origin and dispersion of indigenous pottery, along with TL ages (Bailiff 2017). This will reinforce data pointing toward pottery dispersion throughout the territory during the late Holocene.

Finally, we would like to emphasize the benefits and importance of the interdisciplinary research, an aspect required in the development of South American Geoarchaeology, where researchers from different disciplines jointly address behavioral and contextual reconstructions. In this regard, the case study presented integrates the data generated from different disciplines, contributing to the knowledge about the temporality of the use of space by pottery populations and the generation of geomagnetic field data. It is important to emphasize that the particular strengths of the study method are the applicability in cooked clays, baked stone, and ceramic materials, which survive well in the archaeological record. According to Batt et al. (2017), the method may be more accurate than other techniques for certain time periods, for example, in periods when radiocarbon dating have major errors.

Acknowledgements We appreciate the support of CONACyT's project 258270. We would also like to thank Gabriel Arena and Carola Castiñeira for her comments and suggestions.

References

Aidona E, Kondopoulou D (2012) First archaeomagnetic results and dating of Neolithic structures in northern Greece. Stud Geophys Geod 56:827–844

Austral A (1977) Arqueología de urgencia en el yacimiento de Bañadero, Departamento de Salto. Uruguay. In: Seminario sobre Medio Ambiente y Represas, vol 2. Facultad de Humanidades y Ciencias, Universidad de la República, Montevideo, pp 3–20

Baeza J, Taddei A, Femenías J, Rodríguez O, Melgar W, Díaz A, Fornaro M (1977) Investigaciones Arqueológicas en el Área de Salto Grande: Tres Primeros Radiocarbonos. In: V Encuentro de Arqueología del Litoral. Ministerio de Educación y Cultura and Intendencia Municipal de Río Negro, Río Negro, pp 67–88

Bailiff IK (2017) Luminescence dating of pottery and bricks. In: Gilbert AS (ed) Encyclopedia of geoarchaeology. Encyclopedia of earth sciences series. Springer Dordrecht, Heidelberg, New York, pp 494–498

Batt CM, Brown MC, Clelland SJ, Korte M, Linford P, Outram Z (2017) Advances in archaeomagnetic dating in Britain: new data, new approaches and a new calibration curve. J Archaeol Sci 85:66–82

Beovide L, Campos S (2014) Interacciones entre las sociedades y las plantas durante el Holoceno medio-tardío en el área septentrional del Río de la Plata (Uruguay). Rev Esp Antropol Am 44(2):575–601
Beovide L, Malán M (2009) Puerto La Tuna: Nuevos datos para un viejo debate. In: La arqueología como profesión: los primeros 30 años. Asociación Uruguaya de Arqueología, Montevideo, pp 97–110
Bracco R (1991) Dataciones ^{14}C en sitios con elevación. Rev Antropol 1(1):11–17
Bracco R, Ures C (1999) Ritmos y dinámica constructiva de las estructuras monticulares. Sector sur de la cuenca de la Laguna Merín-Uruguay. In: López JM, Sans M (eds) Arqueología y Bioantropología de las Tierras Bajas. Facultad de Humanidades y Ciencias de la Educación, Departamento de Publicaciones de la Universidad de la República, Montevideo, pp 13–33
Bracco R, Cabrera L, López JM (2000a) La Prehistoria de las Tierras Bajas de la Cuenca de la Laguna Merín. In: Durán A, Bracco R (eds) Arqueología de las Tierras Bajas. Ministerio de Educación y Cultura, Montevideo, pp 13–38
Bracco R, Fregeiro MI, Panarello H, Odino R, Souto B (2000b) Dieta, modos de producción de alimentos y complejidad. In: Durán A, Bracco R (eds) Arqueología de las Tierras Bajas. Ministerio de Educación y Cultura, Montevideo, pp 227–248
Capdepont I (2001) Análisis cerámico de la región este de Uruguay. In: Arqueología Uruguaya hacia el fin del milenio. Gráficos del Sur, Montevideo, pp 377–386
Capdepont I (2013) Arqueología de sociedades indígenas del litoral del río Uruguay. Paisajes y ocupaciones humanas. PUBLICIA, Germany
Capdepont I, Castillo A (2001) Caracterización cerámica para una interpretación antropológica. In: Arqueología Uruguaya hacia el fin del milenio. Gráficos del Sur, Montevideo, pp 120–127
Capdepont I, del Puerto L, Ramírez A (2013) Potenciales fuentes de aprovisionamiento para la manufactura cerámica: sitio Guayacas-Paysandú, Uruguay. Rev Mus Plata Sección Antropol 13(87):2–17
Capdepont I, Castiñeira C, del Puerto L, Fernández G (2016) Desarrollo de las ocupaciones humanas durante el Holoceno en la cuenca de la Laguna de Castillos (Uruguay): síntesis y actualización de las investigaciones arqueológicas. TESSITURAS 4(1):53–93
Capdepont I, Castiñeira C, Blasi A, del Puerto L (2017) Conectividad social durante el Holoceno tardío en el paisaje arqueológico del litoral oriental del bajo río Uruguay. PESQUISAS Antropol 73:93–120
Carmona JM, Osete L, Campuzano Saioa A, Pavón-Carrasco FJ, Larrazabal J (2014) El Arqueomagnetismo como técnica de datación: Ejemplo de estudio en Fonte do Milho (Peso da Régua, Portugal), un poblado de la Segunda Edad del Hierro e inicios del Imperio. In: Actas de las IV Jornadas de jóvenes investigadores del valle del Duero. Glyphos Publicaciones, Segovia, pp 175–186
Castillo A (2001) Arqueología de Río Negro, Uruguay. Rev do CEPA 24(31):71–94
Castillo A (2004) Excavación y Museo: Profundizando en el conocimiento de los grupos ceramistas del litoral (Río Negro-Uruguay). In: Beovide L, Barreto I, Curbelo C (eds) La Arqueología Uruguaya ante los desafíos del Nuevo Siglo. CD-ROM, Montevideo. ISBN 9974-7811-0-8
Castiñeira C, Panario D, Bracco R, Gutiérrez O (2010) Concheros en la costa atlántica uruguaya y su vinculación con la dinámica litoral. In: Arqueología Argentina en los inicios de un Nuevo Siglo. Laborde Libros Editor, Rosario, pp 635–643
Chauvin A, García A, Lanos P, Laubenheimer F (2000) Paleointensity of the geomagnetic field recovered on archaeomagnetic sites from France. Phys Earth Planet Inter 120:111–136
Cobas I, Prieto MP (1998) Regularidades espaciales en la cultura material: la cerámica de la Edad de Bronce y la Edad de Hierro en Galicia. Gall 17:151–175
Coe RS (1967) Paleointensities of the Earth's magnetic field determined from tertiary and quaternary rocks. J Geophys Res 72:3247–3262
Coe R, Grommé S, Mankinen EA (1978) Geomagnetic paleointensities from radiocarbon-dated lava flows on Hawaii and the question of the Pacific non dipole low. J Geophys Res 83:1740–1756
Consens M (2007) Arte prehistórico en Uruguay. Torre del Vigía Ediciones, Montevideo

del Puerto L, García-Rodríguez F, Bracco R, Inda H, Capdepont I, Castiñeira C, Blasi A, Fort H, Mazzeo N (2012) Historia ambiental y dinámica cultural para el Holoceno medio y tardío en el este del Uruguay. In: En clave inter. Procesos, contextos y resultados del trabajo interdisciplinario. Universidad de la República, Montevideo, pp 99–109

del Puerto L, Capdepont I, Inda H (2016) Paleoetnobotánica y subsistencia de los constructores de Cerritos del Holoceno tardío en el Este del Uruguay: análisis fitolítico en sedimentos y artefactos arqueológicos. TESSITURAS 4(1):117–160

Díaz A (1977) Arqueología de Salto Grande: secuencia cultural resultante de las investigaciones realizadas en la Isla de Arriba y del Medio (Uruguay). In: V Encuentro de Arqueología del Litoral. Ministerio de Educación y Cultura e Intendencia Municipal de Río Negro, Río Negro, pp 155–164

Díaz A, Fornaro M (1977) Intento de sistematización de las Modalidades Alfareras del Litoral Uruguayo. In: V Encuentro de Arqueología del Litoral. Ministerio de Educación y Cultura e Intendencia Municipal de Río Negro, Río Negro, pp 165–174

Durán A (1990) Prehistoria del Uruguay. Clasificación de las formas de los recipientes cerámicos. Dédalo 28:109–145

Ech-Chakrouni S, Hus J, Spassov S (2013) Constraints of archaeomagnetic dating and field intensity determinations in three ancient tile kilns in Belgium. Stud Geophys Geod 57:585–604

Farías M, Femenías J, Iriarte J, Florines A (2001) Aproximación Arqueológica del área deltaica del Río Negro. In: Arqueología Uruguaya hacia el fin del milenio. Gráficos del Sur, Montevideo, pp 125–134

Femenías J (1990) Cerámica de los cerritos de noreste de Uruguay y sureste de Brasil (discusión de los modelos resultantes). Rev do CEPA 17(20):153–156

Genevey A, Gallet Y (2002) Intensity of the geomagnetic field in Western Europe over the past 2000 years: New data from ancient French pottery. J Geophys Res 107:1–18

Guidón N (1989) Los sitios de las islas. El sitio Y58. In: Misión de Rescate Arqueológico de Salto Grande, Tomo II. Ministerio de Educación y Cultura, Montevideo, pp 433–572

Hartmann GA, Afonso MC, Tindade RI (2007) Arqueomagnetismo e datação arqueomagnética: princípios y métodos. Rev do Mus Arqueol e Etnol 17:445–459

Henrickson EF, MacDonald M (1983) Ceramic form and function: an ethnographic search and archaeological application. Am Anthropol 85:630–645

Hilbert K (1991) Aspectos de la Arqueología en Uruguay. Ava Materialien 44. Verlag Philipp Von Zabern. Mainz am Rhein, Germany

Hogg AG, Hua Q, Blackwell PG, Buck CE, Guilderson TP, Heaton TJ, Niu M, Palmer JG, Reimer PJ, Reimer RW, Turney CS, Zimmerman S (2013) SHCal13 southern hemisphere calibration, 0–50,000 years cal BP. Radiocarb 55(4):1889–1903

Hus J, Geeraerts R, Spassov S (2003) Archaeomagnetism and archaeomagnetic dating. Institut Royal Météorologique de Belgique, Belgique

Kolb CC (2017) Ceramics. In: Gilbert AS (ed) Encyclopedia of geoarchaeology. Encyclopedia of earth sciences series. Springer Dordrecht, Heidelberg, New York, pp 118–128

Langouet L, Goulpeau L (1984) La datation archaeomagnétique du temple du Haut-Bécherel a Corseul. Rev Arch Oueste 1:85–88

Lengyel S (2017) Archaeomagnetic dating. In: Gilbert AS (ed) Encyclopedia of geoarchaeology. Encyclopedia of earth sciences series. Springer Dordrecht, Heidelberg, New York, pp 39–46

Leonhardt R (2006) Analyzing rock magnetic measurements: the Rock-MagAnalyzer 1.0 software. Comput Geosci 32:1420–1431

Leonhardt R, Heunemann C, Krása D (2004) Analyzing absolute paleointensity determinations: acceptance criteria and the software ThellierTool4.0. Geochem Geophys Geosyst 5(12):1–11

López JM (2000) Trabajos en tierra y complejidad cultural en las tierras bajas del Rincón de los Indios. In: Durán A, Bracco R (eds) Arqueología de las Tierras Bajas. Ministerio de Educación y Cultura, Montevideo, pp 275–281

López JM (2001) Las estructuras tumulares (cerritos) del litoral atlántico uruguayo. Lat Am Antiq 12(3):231–255

López JM, Castiñeira C (2001) Estructura de sitio y patrón de asentamiento en la Laguna Negra Rocha. In: Arqueología Uruguaya hacia el fin del milenio. Gráficos del Sur, Montevideo, pp 147–162

López JM, Villarmarzo E, Brum L (2009) Análisis de secciones y plantas arqueológicas del Sitio La Esmeralda (Rocha, Uruguay). In: Beovide L, Erchini C, Figueiro G (eds) La arqueología como profesión: los primeros 30 años. CD-Rom, Montevideo, pp 218–229

López JM, Dabezies M, Capdepont I (2014) La gestión de los recursos vegetales en las poblaciones prehistóricas de las tierras bajas del sureste del Uruguay: un abordaje multidisciplinar. Lat Am Antiq 25(3):256–277

McCabe C, Jackson M, Ellwood B (1985) Magnetic anisotropy in the Trenton limestone: results of a new technique, anisotropy of anhysteric susceptibility. Geophys Res Lett 12:333–336

Morales J, Goguitchaichvili A, Aguilar-Reyes B, Pineda-Duran M, Camps P, Carvallo C, Calvo-Rathert M (2011) Are ceramics and bricks reliable absolute geomagnetic intensity carriers? Phys Earth Planet Int 187:310–321

Morales J, Goguitchaichvili A, Aguilar B, Pineda M, Carvallo C, Baramendi L, Hernandez G, Oliveros A (2012) A detailed rock-magnetic and archaeointensity investigation on some pottery relics and burned floors from Tzintzuntzan archaeological site, Western Mesoamerica. Geoarchaeology 27(6):521–537

Naue G (1973) Datos sobre o estudo dos cerritos na aera meridional da Lagoa dos Patos, Río Grande, R.G.S. Veritas 71–73

Novillo Verdugo MA, Esparza López R, Goguitchaichvili A, Morales J, Cervantes Solano M (2016) Datación de tinajas arqueológicas del sitio Guachimontones, occidente de Mesoamérica. Arqueol Iberoam 29:9–14

Orton C, Tyers P, Vince A (1997) La cerámica en Arqueología. Crítica, Barcelona

Outram Z, Batt CM (2010) Dating at Old Scatness. In: Dockrill SJ, Bond JM, Brown LD, Turner VE, Bashford DJ, Cussans JE, Nicholson RA (eds) Excavations at Old Scatness, Shetland, vol 1: The Pictish Village and Viking Settlement. Shetland Heritage Publications, Lerwick, pp 99–130

Parés JM (2017) Paleomagnetism. In: Gilbert AS (ed) Encyclopedia of geoarchaeology. Encyclopedia of earth sciences series. Springer Dordrecht, Heidelberg, New York, pp 601–607

Pavón-Carrasco FJ, Rodriguez-Gonzalez J, Osete ML, Torta J (2011) A Matlab tool for archaeomagnetic dating. J Archaeol Sci 38(2):408–419

Philippi RA (1847) Testaceorum novorum centuria. Z für Malakozoologie 4(5):84–96

Pintos S, Capdepont I (2001) Arqueología en la Cuenca de la Laguna de Castillos: apuntes sobre complejidad cultural en sociedades cazadoras recolectoras del Este del Uruguay. ArqueoWeb 3 (2). http://www.ucm.es/info/arqueoweb. Accessed 13 octubre 2017

Politis G, Bonomo M (2012) La entidad arqueológica Goya-Malabrigo (ríos Paraná y Uruguay) y su filiación Arawak. Rev Arqueol 25(1):10–46

Prevosti M, Casas L, Roig Pérez JF, Fouzai B, Álvarez A, Pitarch A (2013) Archaeological and archaeomagnetic dating at a site from the ager Tarraconensis (Tarragona, Spain): El Vila-sec Roman pottery. J Archaeol Sci 40:2686–2701

Rice PM (1987) Pottery analysis: a sourcebook. University of Chicago Press, Chicago

Riisager P, Riisager J (2001) Detecting multidomain magnetic grains in Thellier palaeointensity experiments. Phys Earth Planet Inter 125:111–117

Schmitz PI (1968) Grandes complexos de cerâmica indígena no Sul do Brasil. Pesqui Antropol 18:127–140

Schmitz PI (1991) Migrantes da Amazonia: a Tradicao Tupiguarani. In: Arqueologia y Pre-Historia do Rio Grande do Sul, Editora Mercado Aberto Ltda., Porto Alegre, pp 295–330

Schmitz PI, Naue G, Basile Becker I (1991) Os aterros dos campos do Sul: a Tradiçao Vieira. In: Arqueologia y Pre-Historia do Rio Grande do Sul. Editora Mercado Aberto Ltda., Porto Alegre, pp 221–250

Selkin PA, Gee JS, Tauxe L, Meurer WP, Newell AJ (2000) The effect of remanence anisotropy on paleointensity estimates: a case study from the Archean Stillwater Complex. Earth Planet Sci Lett 183:403–416

Serrano A (1972) Líneas fundamentales de la arqueología del litoral. Universidad Nacional de Córdoba, Instituto de Antropología, Dirección General de Publicaciones, Córdoba
Shennan S (1992) Arqueología Cuantitativa. Editorial Crítica, Barcelona
Shepard A (1954) Ceramics for the archaeologist. Publication 09. Carnegie Institution of Washington, Washington, DC
Steward JH (1946) Handbook of South American Indians, vol 1: The Marginal Tribes. Smithsonian Institution—Bureau of American Ethnology, Washington, DC, Bulletin 143
Thellier E, Thellier O (1959) Sur l'intensité du champ magnétique terrestre dans le passé historique et géologique. Ann Géophys 15:285–376

Site Formation Processes Since the Early Holocene in a Planosol Sequence from the Laguna Negra Basin

Alfonso Machado Arnaud, Gustavo Piñeiro Barceló and José López Mazz

Abstract The low lands in southeastern Uruguay have witnessed the beginning of the traditions of mound builders on this side of the world. However, evidence of the first settlers in the area is scarce. In this paper, a site record is presented which dates to the early Holocene in the region, which could significantly change our perspective on the settlement process and further cultural developments. Based on the granulometric, color, carbonate, and organic matter values for the sediments in the soil stratigraphy, it is observed that the formation processes occurred toward the end of the loess sedimentation, which is typical of the late Pleistocene in the Pampas region. The unit corresponding to the early Holocene is in the contact layer between the Pampas silts and the overlying sands, creating a pale gray, sandy-silt lens with a high density of archaeological materials. It is concluded that, until the late Pleistocene, the landscape consisted mainly of silty soils of aeolian origin, followed by a sudden, net, nonerosive change which can be observed if compared to the later units, characterized by a sequence of upward-fining sand layers. The deposit process occurred along with the sedimentation of a granulometric fraction which coexists with the main sandy fraction throughout the profile, together with an abundant cultural record. The contrast of hydraulic transmissivity between the loess and bimodal sands has contributed to the development of an eluvial horizon and consequently, of a typical planosol. The underlying unit contains abundant lithic archaeological materials, which represent the first settlements in the current Atlantic plains of Uruguay during the late Pleistocene sedimentation process and prior to the paleo-environmental change, which resulted from an increase in humidity and temperature in the early Holocene.

Keywords Geoarchaeology · Early Holocene · Planosol

A. M. Arnaud (✉) · G. P. Barceló
Facultad de Ciencias, Universidad de la República, Iguá 4225, 11400 Montevideo, Uruguay
e-mail: alfo1977@gmail.com

J. L. Mazz
Facultad de Humanidades y Ciencias de la Educación, Universidad de la República, Av Uruguay 1695, 11200 Montevideo, Uruguay

H. Inda Ferrero and F. García Rodríguez (eds.), *Advances in Coastal Geoarchaeology in Latin America*, The Latin American Studies Book Series,
https://doi.org/10.1007/978-3-030-17828-4_10

The Biophysical Environment

Climate and Vegetation

The basin of Laguna Negra is located between 31°30′–34°30′S and 52°–55° W, comprising an area of approximately 54,000 Km^2. Uruguay's climate is mild with a mean annual precipitation of 1000 mm, an average temperature of 17 °C, and a mean wind speed of 7 m/s (Prohaska 1976). According to Cabrera and Willink (1973) and Morrone (2001, 2006), the country is located in the Pampas Biogeographic Province.

Soils

There is a correspondence between the geomorphology and the intrazonal soils. The higher areas of the basins of Laguna Negra and Laguna Merin present well-drained soils, while the middle and lower sections are mainly hydromorphic soils. Udolls are the zonal and the most common regional soil in the basin and predominate in the higher areas of the basin, while in the gentle slopes and the lowlands gleysols, pseudogleys, natrudolls, and planosols are most commonly found. An exhaustive revision of the general features and the genesis of the soils in the region was carried out by Sombroek et al. (1969).

The surroundings of the archaeological site of Los Indios are at the point of contact of both abovementioned soils (Fig. 1), with well-drained soils in the higher areas and hydromorphic soils in the low wetlands, characterized by the predominance of gleysols and histosols. According to Altamirano et al. (1976), these soils often lack an argiluvi horizon and present a histic horizon of less than 30 cm of thickness. However, soils with peaty horizons of several meters of thickness have been described in the basin (Goso 1985). There is a constant presence of alkaline soils in the middle sections of the highlands and far from water courses, which was early interpreted by Sombroek et al. (1969) as the result of the presence of nearby marine conditions.

Geology and Geomorphology

The superficial geology at the basins of Laguna Negra and Laguna Merin is complex. The basement of the area is made up of the Rocha Formation (Precambrian) and granitoids with varied compositions, granitic gneiss, and migmatites, intercalated with mafic rocks, calcium silicates, and quartzite. The Rocha Formation is composed by low-grade metamorphic rocks generally developed toward a north-northwest direction, stretching from the cities of Rocha and La Paloma up to Laguna Negra which to a large extent, determined the structure and direction of the Atlantic coast faced with the eastern oceanic winds. The Santa Teresa granite (Ediacaran–Cambrian) is one of the

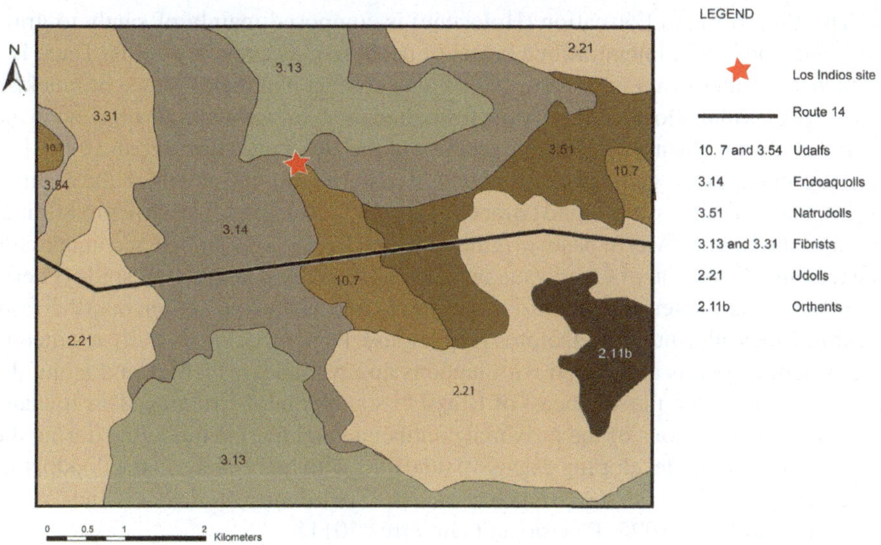

Fig. 1 Soils of the area *Source* Ministerio de Ganadería Agricultura y Pesca/Dirección General de Recursos Naturales/CONEAT (1979)

largest intrusions in Uruguay (Hasui et al. 1975). The Chuy formation (Pleistocene) is found in the regional sedimentary sequence since the late Pleistocene, made up of fine to medium sand and clay–sand sediments of white, yellowish, and reddish colors, of littoral marine sedimentation, alternating between siltstones and claystones and coarse sandstones (Goso 1985). The Libertad Formation (Pleistocene) is characterized by mud and loess with varying percentages of brown to reddish brown sand and clay, with calcium carbonate concretions of different shapes. The sedimentation environment is Continental-peridesertic, with periods of greater humidity that enabled the accumulation of fine material by aeolian transport and its removal together with the alteration mantles by runoff, mass wasting, and solifluction (Preciozzi et al. 1985).

The Dolores Formation (Goso 1985) is composed of clay–silt and silt–clay deposits, forming muds, and very fine clay sands with varying contents of sand and floating gravel, having a massive matrix with varying compaction degrees depending on the clay content. Calcium carbonate is present and distributed in a disperse, pulverulent, or concretional manner. It is primarily brownish and subordinately gray to greenish gray, of continental sedimentation related to fine-sediment aeolian phenomena (i.e., loess-like sediments) from the arid areas located southwest of the area of study, as well as to colluvium phenomena with the formation of mudflows, which were re-deposited in topographically lower areas (Preciozzi et al. 1985). At its top, it sometimes presents archaeological materials associated with the first settlement of the region (López Mazz et al. 2009; Meneghin 2004, 2006; Nami 2013; Suárez 2011; among others).

The Villa Soriano Formation (Holocene) is composed mainly of sandy to gravelly sediments, with interstratified layers of pebbles, clays, and gray silts. There is a sequence of sand strata with more or less clay content and the presence of brackish water species that allow to identify the formation as an ingressive environment on the continent during the maximum transgressive of the Holocene (Bossi et al. 1998). It is also presented as the geological formation in eastern Uruguay on which the cultures of "mound builders" were found (Bracco et al. 1992, 2000a, b). The structural framework of the Laguna Merin basin originated in the Cretaceous graben accompanying the tectonic distension of Gondwana, which was filled by potent basalt spills (Puerto Gómez formation sensu Preciozzi et al. 1985) and Cenozoic sediments (the Fray Bentos, Libertad, Chuy, and Dolores formations). Its neotectonic activity resulted in the system of plains associated with lagoons and wetlands which extend along the Atlantic coast, from the southeast of Brazil to south-central Uruguay. This lowland system had oscillations of the proximal continental and marine modeling during the Quaternary Period, developing extensive tidal flats with ingressions in the floodplains of the current water courses (Bracco et al. 2012; del Puerto et al. 2011; Inda 2011; Montaña and Bossi 1995; Panario and Gutiérrez 2011).

The geomorphology near the site is characterized by the presence of elevated areas on the surrounding plains. The area known as Potrero Grande Santa Teresa is located to the north and northwest, Angostura ridges to the east and south, Difuntos ridges to the southwest, Bella Vista ridges to the west, and Blanqueada ridges to the northwest (Fig. 2). The highest altitudes reach 170 m above sea level and a road network that makes regional transit possible can be observed in these ridges, where lagoons and wetlands dominate the landscape.

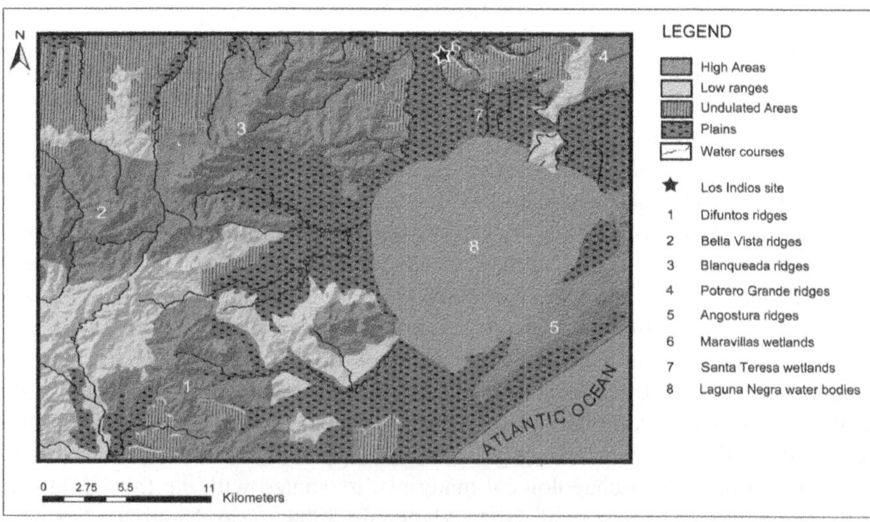

Fig. 2 Geomorphological map of the Laguna Negra basin. Taken from Topographic Sheet C-24 of the Servicio Geográfico Militar Uruguayo; Montaña y Bossi 1995.

Archaeology, Paleo-climates, and Paleo-environments

At a global level, an increase in the environment temperature and humidity was recorded in the Pleistocene/Holocene transition (e.g., Gornitz 2009), which coexisted with the early human settlements of the American continent (Dillehay 2000; Miotti 2006; Politis 1991). The countryside in the territories of Argentina, Uruguay, and southeastern Brazil presents archaeological sites of hunter-gatherer groups, which were typical of the period known as Paleo-Indian (Días 1994, 2012; Favier Dubois 2006; López Mazz 2013; Suárez and López Mazz 2003) in other research conducted in other areas of the globe (Vegas 2007; Nicholson and Flohn 1980; Bar-Matthews et al. 1998).

This climate trend remained until it stabilized during the early Holocene with a warmer, more humid climate than the late Pleistocene (Fig. 3). However, archaeological records for this period are scarce, both in the area studied and in the entire Uruguayan territory, given that the early coastal sites would be at lower levels than the current levels of the sea.

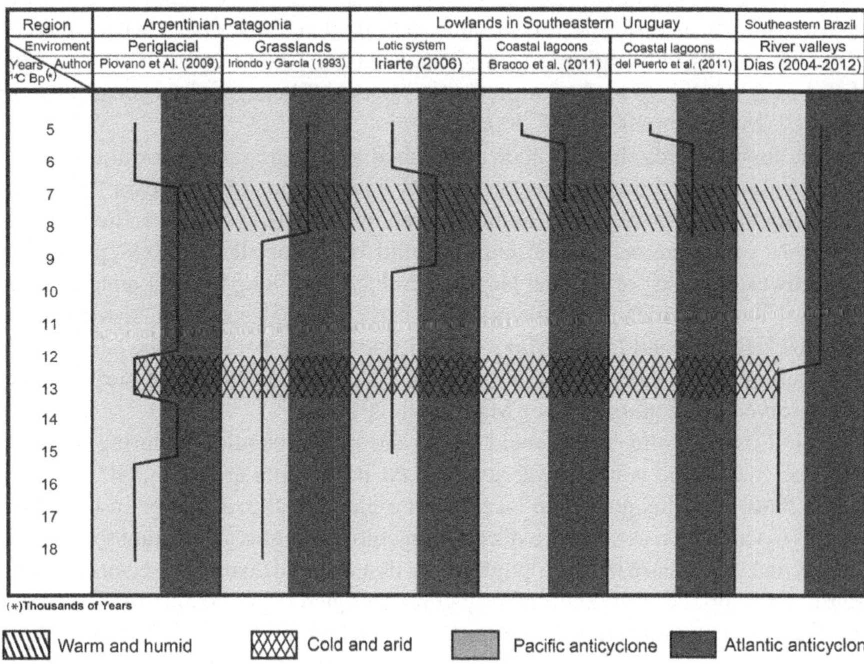

Fig. 3 Paleoclimatic table of the Pleistocene/Holocene and early Holocene transition in the American southern cone (Piovano et al. 2009; Iriondo and García 1993; Iriarte 2006; Bracco et al. 2011; del Puerto et al. 2011; Días 2012)

North and East of the Uruguayan Territory

Northwest of the Laguna Merin basin, in the basaltic slope of the Uruguayan territory, the geological unit corresponding to the late Pleistocene is defined as the Sopas-Dolores Formation (sensu Goso 1965; Panario and Gutiérrez 1999). This formation shows a gradual increase of humidity and temperature in the environment (Ubilla 1999, Castiñeira 2008) where the local Young paleofauna is described, showing a significant presence of megafaunal species, ^{14}C dated to 11,650 years BP by Ubilla and Perea (1999). This training also has a maximum date of ca, 45,000 BP year (Bracco et al. 2001).

In the middle section of the Uruguay River, human occupations were identified in a paleo-wetland ^{14}C dated to ca. 11,200 years BP (Guidon 1989). In the area, two stratigraphic groups are identified, one corresponding to the Pleistocene/Holocene transition as heavy alluvium and the other to the Holocene as a light alluvium (Piñeiro et al. 1999). In the Cuareim river basin, the site of Pay Paso presents remains of early settlements in a fluvial sequence developed from the rocky basement between ^{14}C 11,000 and 7,500 years BP (Austral 1995; Piñeiro and Suárez 2012).

The oldest archaeological sites closest to the study area are found on Cerro Los Burros, a hill located on the transitional Atlantic–Estuarine coast, where there is a Unit which was radiocarbon dated to between ca. 10,690 and 11,690 years BP, defined as a "tenacious rock that presents cracks and hexagonal fissures inside its structure" (Meneghin 2004, p. 7, 2006, p. 3).

In the lowland area, Iriarte (2006) described a stratigraphic sector at the site of Los Ajos dating to the late Pleistocene, ^{14}C dated to ca. 14,810 years BP, where Chroloideae sp. from very arid climates predominate. However, the site does not show signs of human occupation until the mid-Holocene (Iriarte 2006, p. 26). On the northwestern coast of Laguna Negra, a collection of long bifacial materials was recovered, among which projectile points of early typology stand out, also known as "fish tails" (Bracco and López Mazz 1992; López Mazz 2013). As the site is under water, neither a stratified context nor absolute dating of the archaeological set have been observed or obtained (López Mazz et al. 2010).

At the Los Ajos site, it presented warm and humid conditions during the early Holocene, where the wetland was established in the area around 8,840 BP years (Iriarte 2006). For this period an increase in organic matter, a change in local flora from C3 to C4 plants, as well as a decrease in chloridoid grasses tolerant to very arid climates and an increase in Myriophyllum sp. as a wetland taxon have been described (Iriarte 2006, pp. 27–28). Sedimentary cores sampled at the Santa Teresa wetland presented, in a stratum prior to the middle Holocene, redeposited sponge spicules and marine foraminifera (Elphidium sp.) (Blasi et al. 2005). There is no presence of diatoms and phytoliths, and organic matter as well as calcium carbonates are minimally represented (Inda 2011; Blasi et al. 2005). According to Bracco et al. (2005), this Unit correlates with the Villa Soriano formation (sensu Preciozzi et al. 1985).

The Archaeological Site of Los Indios

The geomorphological environment of this site presents two alluvial plains that form facing peninsulas at 2 km distance; to the east, the site of Los Indios, and to the west, the site of Peninsula Oeste. This forms a gap that allows the animal flow of the lower area, which is dominated by the dynamics of the Los Indios stream, the wetlands of Bañado de Santa Teresa and Las Maravillas, and Laguna Negra.

Even though arrays of mound-shaped structures from the late Holocene, known as "Indian mounds" (radiocarbon dated to between 2,900 and 770 years BP), stand out at the site, an earlier archaeological component was recorded dating to the early Holocene in the interval between the 8,500 and 7,100-year BP (López Mazz 2013). It is an unusual record in the Uruguayan territory as well as in the whole of the Laguna Merin basin. Having the occupation of the site been so prolonged, the analyses carried out on the lithic material show that the relocation movements of human groups decreased over time (Gascue et al. 2009).

The beginning of the settlement is represented in Stratification Unit (SU) 04 of the fifth excavation that is at the top of the Dolores Formation; it is brown with chestnut veins and compaction cracks, and contains lithic and faunal archaeological materials (López Mazz 2013). On this SU there is SU03, AMS dated to the early Holocene ca. 8,560 and 7,100 years BP and described as light brown, sandy-silt soil with gravel and fine gravel, which contains abundant lithic materials and a combustion structure with small, very eroded remains of fauna, with no signs of bearing, belonging to small and mid-sized mammals, calcareous shellfish, fish and birds (Sotelo and López Mazz 2015). Regarding this SU, there have been different interpretations about its formation, as may be the influence of ash from local fires as well as volcanic ash, given its particular clear colors (Fig. 4).

By means of the many datings obtained from the excavations of the site, an archaeological and sedimentary hiatus is observed in the transition from the early Holocene to the mid-Holocene and for almost all of the latter period (ca. 7,100–2,900 years BP) (López Mazz 2001, 2013). SU02 is dated to the late Holocene and is correlated to the construction period of the Indian mounds. It is clayey silt with gravel and fine gravel, and contains lithic and ceramic materials. SU01 is the most superficial unit: dark brown, gravelly silt with lithic archaeological material (López Mazz 2001).

Methodology and Research Strategy

The present work aims to delimit the culture material and the environmental indicators that took place in the formation of the early strata of the site, according to the works carried out in the fifth expansion of the excavation. Samples of sediments from each SU were extracted and physical textural and geochemical analyses were conducted in order to discuss the paleo-environmental and paleo-climatic conditions in which these prehistoric societies coexisted in the early Holocene, in a period comprised between the Paleo-Indian and the Formative periods.

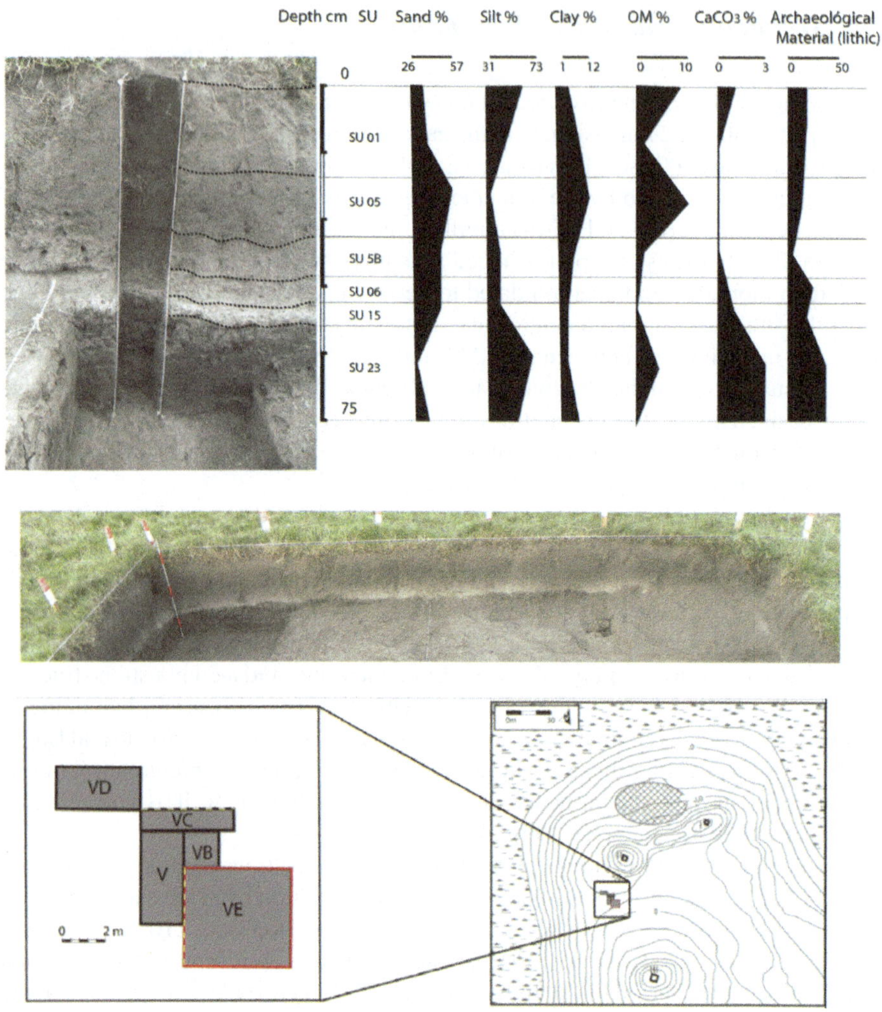

Fig. 4 Los Indios site, excavation, stratigraphic profile, and analytical values (modified from López Mazz and Gianotti 2001)

Sediment Sampling and Analysis

During the fifth expansion of the fifth excavation at the site, 15 samples of a sedimentary column were collected at 5-cm intervals, 7 of which were selected, corresponding to the 6 SUs recorded in layers covering the entire area. In situ, the profile was described from base to top as coarsening upward, with a very heavy base unit that presents compaction cracks (SU23), followed by a particularly clear unit in comparison to the rest (SU15), from where units of brown to dark colors, intruded by roots and abundant archaeological materials, were developed (EU 06, 5B, 05, and 01).

The method used for the granulometric analysis included two techniques for separating the sediment fractions by grain size according to the Udden–Wentworth scale (1922). Psephites were separated by wet sieving in mesh openings of 1Ø–4Ø. Likewise, due to the differential capacity of objects to remain suspended in water according to Stokes's Law, pelitic fractions were separated using the International Pipette Method, which consists of taking a subsample with a pipette at a given depth and time, resulting in the elimination by sedimentation of all particles larger than a certain diameter (Dewis and Freitas 1970).

Colorimetric observation was made with the samples in dry conditions, with natural light and using the Munsell Soil Color Chart (1992).

To calculate the contents of organic matter and carbonate, the loss on ignition (LOI) method was used as described by Heiri et al. (2001). The general principle is based on the fact that both organic matter and carbonate are temporarily transformed into carbon dioxide (CO_2) gas at a specified temperature. At the end of this process, the mass loss of the sediment is redetermined, having an equivalence for each component (Heiri et al. 2001).

Mineralogical analyses were conducted using a Leitz Wetzlar Laborlux 12 Pol S 50-500X polarizing microscope, on preparations immersed in Canada balsam ($n = 1.55$).

Results

Grain Size, Colorimetry, Organic Matter, Carbonate, Mineralogy

The process of formation of archaeological soil begins in the stratigraphy dominated by the pelites, where the silt fraction stands out, together with a secondary mode in coarse sands in the upper SUs (5B, 05, and 01). The clay fraction is concentrated at the base of the sequence (SU23), as well as in SUs 5B and 05. The sands have an upward grading, while the silts are consistent throughout the profile, with the highest concentration in SU23 and the lowest in SE5B. The granulometric evolution is linear except in SU5B, which shows an increase in sands and clays and the consequent decrease in silt compared to the rest of the column. The SUs are characterized as ascending sandy–silt to silty–sand deposits, depending on the relative proportion of mineral grain fractions (sensu Folk and Ward 1957).

The coloration of the soil changes from grayish brown to darker, almost black shades for the upper SUs, with lighter shades of gray (SU15 and SU06) and particularly light grays for SU15 (Munsell 1992).

The organic matter content is null in the lower fraction of SU23 as well as in SU15, with a low percentage in SU06 and SU05 (2%), a medium one in the upper section of SU23 (5%), but high in SU5B (10%) and SU01 (9%). Carbonates have higher contents in the lowest SU (SU23 = 3%), while presenting a 1% content in SU15 and SU01, and being null in the remaining SUs (SU06, SU5B, and SU05) (Table 1).

The mineralogy of the sandy fraction is mainly quartz (75–85%) followed by potassium feldspar and plagioclase (7–15%), in general with alteration along the cleavages, with the presence of varnish of iron oxides and manganese. As accessory minerals, zircons of different morphologies are observed together with partially altered and fresh biotite in similar proportions, opaque minerals (mainly magnetite), garnet, tourmaline and muscovite in decreasing proportions. The significant differences in the mineralogy of the samples throughout the profile refer to the relative percentages of quartz and feldspar, being the former more frequent in the finer sand samples.

In the silty fraction, lower percentages of quartz and feldspars were observed, while there is an increase in the relative concentrations of opal phytoliths and pollen grains of diverse characters, which are not addressed in this analysis. No amorphous particles of volcanic ash were observed.

Description of Early Site Units

According to the field record, the stratigraphic profile of the archaeological excavation is composed of five overlapping SUs of deposits in layers. SU23 is undulating at the top, forming basins which are filled by the upper SUs. It is more than 1 m thick and has a sandy-silt texture with a greater amount of clay and sand in some sections, a very dark grayish-brown to gray color, high tenacity in dry conditions and high plasticity in wet conditions. It contains the largest amount of clay throughout the stratigraphic column. However, the silt fraction predominates, followed by very fine sands. The organic matter increases from 0 to 5%, while the carbonates are mostly represented in this SU if compared to the rest of the profile. These appear pulverized and in crusts that cover small bone and malacological remains as well as enamel from teeth. The matrix contains gravel and coarse sands of different geneses, mostly composed of aggregates of feldspars (plagioclase and potassium), quartz (white, gray and pink, among others) and micas (biotites and muscovites), with no signs of bearing. On the other hand, there is a variety of quartz with signs of bearing by river transport. In this SU, there appear compaction cracks filled by a black sediment with a thicker texture which contains abundant lithic archaeological materials. These intrusions that differ from their matrix were first described as SU34. SU15 has a sandy-silt granulometry, although it has more sand than SU23's average. It is a layer of 4–7 cm of thickness, with undulating features at the base and top, a very light gray color, no presence of organic matter and carbonate, and with abundant lithic and bone archaeological material.

SU06 has a sandy-silt texture, and it contains less pelites and more psephites than SU15. It is a light brown to gray color, degraded between the underlying SU15

Table 1 Grain size analysis, colorimetry, carbonate percentage, and organic matter

Sample	Depth (cm)	SU	Sand (%)	Silt (%)	Clay (%)	Folk and Ward classification (1957)	Munsell color (1992)	$CaCO_3$ (%)	OM (%)
14	00–05	01	34	64	1	Silty sand	10YR 2/1 black	1	9
12	10–15	05	36	55	9	Sandy silt	10YR 2/1 black	0	2
10	20–25	5B	57	31	12	Silty sand	10YR 3/1 very dark gray	0	10
08	30–35	06	48	50	2	Sandy silt	10YR 6/2 light brownish gray	0	2
07	35–40	15	49	50	1	Sandy silt	10YR 7/1 light gray	1	0
05	45–50	23	26	73	1	Sandy silt	10YR 5/1 gray	3	5
01	65–70	23	31	60	9	Sandy silt	10YR 3/2 very dark grayish-brown	3	0

(light gray) and the overlying SU5B (very dark gray). Abundant lithic and bone archaeological materials are present.

Discussion and Conclusions

The site provides new data on the climatic and environmental conditions during the Pleistocene/Holocene transition and in the early Holocene. SU23, although it has no absolute datings, it is in a lower position than the SU dated to the early Holocene (SU03 at the 5A and 5C excavations) (sensu López Mazz 2013). This SU presents abundant archaeological materials similar to the regional typology for the early Holocene and Paleo-Indian periods in the area. In turn, the analyses of grain size, color, and contents of organic matter and carbonate, as well as the archaeological background of the site as described by and the available geological cartography), express that this archaeological context is located at the top of the Dolores Formation, corresponding to the late Pleistocene and its transition into the Holocene.

The floating gravel found in the SU23's matrix has different origins, depending on whether it shows signs of bearing by river transport. Regarding the soil of which the site is composed, SU23 corresponds to a B horizon according to the description of the United Soils Department of America (USDA 2006). These are formed "below an A, E or O horizon, and the original structure of the rock is predominantly destroyed" (USDA 2006, p. 312). Likewise, they may display the following features: (1) illuvial concentration of silicate clay, iron, aluminum, humus, carbonates, gypsum or silica; (2) evidence of carbonate removal or addition; (3) residual concentration of oxides; (4) a block or prismatic granular structure by expansion and compaction; and (5) strong gleization (USDA 2006, p. 312).

SU34 is filling the cracks of SU23 and contains abundant archaeological materials deposited after the formation of SU23, as well as cracking in processes which were contemporary or prior to the upper formation of SU15.

SU15 is comparable to the SU03 of the first three stages of the excavation, dated with two absolute datings to the early Holocene (López Mazz 2013). The sediments are sandy silt (sensu Folk and Ward 1957) and the increase in psephites occurs mostly in coarse sands, which coincides with the early development of the Holocene soil. It does not contain enough ash or carbonate particles that can justify its distinctive light gray color. It corresponds to a typical mineral horizon of a type E soil (eluvial) (USDA 2006). Its main feature is the "loss of silicate clay, iron or aluminium, or a combination of these, where a concentration of sand and silt particles remains…", with little or no organic matter (USDA 2006, p. 311). This process of intense soil leaching occurs due to the filtering capacity of this horizon with respect to the underlying SU23, which is very little permeable. The increase in ambient humidity in the early Holocene, discussed in previous local works (del Puerto et al. 2011; Iriarte 2006), as well as in regional works (Cavallotto et al. 2005; Guidon 1989; Iriondo and García 1993; Piovano et al. 2009; Piñeiro et al. 1999), generated the washing of horizon A that was being formed. As part of a paedogenetic process, it transformed into a mineral horizon due to the loss of organic matter, which is also indicated by a very light color. The taphonomy of the site surveyed in previous works is consistent with these flooding conditions, showing very eroded bones but without signs of bearing according to Sotelo and López Mazz (2015).

SU06 could represent a continuous episode from the deposit of SU15, evidenced in the shift to darker shades, as well as in the grain size, which becomes thicker. This SU should correspond to the early Holocene given its location in the stratigraphy and likewise, it could be chronologically correlated with the basal unit of sediment cores extracted at Bañado Santa Teresa, which were interpreted in previous works by Blasi et al. (2005) and Inda (2011) as representative of brackish flooding conditions.

The analyses of sediments have improved the characterization of the deposits of the early component of the site where, in the case of SU23, the correlation with the top of the Dolores Formation is maintained. The late Pleistocene generated loess deposits and redeposits in the current Uruguayan territory where, at its top, the first human settlements took place according to López Mazz (2013).

The range that characterizes the Pleistocene/Holocene transition in Uruguay is delimited by multidisciplinary studies in which the faunal association representing this period has a chronology comprised between 10,500 and 9,500 years BP for the Dolores Formation, according to Ubilla (1999). The site of Los Indios represents the Pleistocene/Holocene transition in the low lands of eastern Uruguay.

At the site, the early Holocene is represented as an eluvial horizon of the soil with intense human occupation, which may have promoted the compaction of the soil B horizon and consequently, resulted in an increase of the leaching processes on the E sub-horizon.

The late and mid-Holocene periods are represented in the more superficial SUs (05, 5B, and 01), which share grain size characteristics where coarse silts and fine sands predominate, displaying dark colors corresponding to the groups known as "mound builders".

References

Altamirano A, Da Silva H, Durán A, Echeverría A, Panario D, Puentes R (1976) Carta de Reconocimiento de Suelos del Uruguay. Tomo I. Dirección de Suelos y Fertilizantes—Ministerio de Agricultura y. Pesca, Montevideo

Austral A (1995) Los cazadores del sitio estratificado Pay Paso hace 10,000 años. In: Consens M, López Mazz JM, Curbelo C (editors), Arqueología en Uruguay, pp 212–218

Bar-Matthews M, Ayalon A, Kaufman A (1998) Middle to Late Holocene (6,500 Yr. Period) Paleoclimate in the eastern mediterranean region from stable isotopic composition of speleothems from Soreq Cave, Israel. In: Issar AS, Brown N (eds), Water, environment and society in times of climatic change, vol 31. Water Science and Technology Library, Springer, Dordrecht

Blasi A, Castiñeira C, del Puerto L, Inda H, Bracco R, García-Rodríguez F (2005) Sedimentación holocena en los bañados de Santa Teresa y el registro arquelógico, planicie costera del departamento de Rocha, Uruguay. XVI Congreso Geológico Argentino. Simposio Geología del Cuaternario, Paleontología, Geoarqueología, Paleoclimas, Paleoambientes, Buenos Aires, pp 1–5

Bossi J, Ferrando L, Montaño J, Campal N, Morales H, Gancio F, Schipilov A, Daniel D, Sprechmann P (1998) Geocarta—Carta Geológica del Uruguay a Escala 1/500.000 Versión 1.1. Geoeditores S.R.L. Facultad de Agronomía. Montevideo

Bracco R (1992) l Arqueología ¿una ciencia pura y aislada? Arqueología MEC:15–31

Bracco R, López Mazz J (1992) Rescate Arqueológico de la Cuenca de la Laguna Merín: Resultados de la Etapa de Prospección. Arqueología. Ministerio de Educación y Cultura, Montevideo, pp 33–50

Bracco R, Cabrera L, López Mazz J (2000a) La prehistoria de las tierras bajas de la cuenca de la Laguna Merín. In: Durán Coirolo A, Bracco Boksar R (eds) Arqueología de las Tierras Bajas. Americana, Montevideo, pp 13–38

Bracco R, Montaño J, Bossi J, Panarello H, Ures C (2000b) Evolución del humedal y ocupaciones humanas en el sector Sur de la cuenca de la laguna Merin. In: Durán Coirolo A, Bracco Boksar R (eds) Arqueología de las Tierras Bajas. Americana, Montevideo, pp 99–115

Bracco R, Ures C, Ubilla M (2001) Contribución al análisis crítico de la relación espacial y temporal entre material cultural y megafauna (Pleistoceno final y Holoceno) en la cuenca norte de Uruguay. Arqueología uruguaya al fin del milenio. Aua, Montevideo, pp 327–339

Bracco R, Inda H, del Puerto L, Castiñeira C, Sprechmann P, García-Rodríguez F (2005) Relationships between Holocene sea-level variations, trophic development, and climatic change in Negra Lagoon, Southern Uruguay. J Paleolimnol 33:253–263

Bracco R, del Puerto L, Inda H, Capdepont I, Panario D, García-Rodríguez F (2012) Evolución ambiental y constructores de cerritos en la región de India Muerta. Un replanteo. In: III Jornadas del Cenozoico, Asociación Uruguaya de Geología, Montevideo, pp 1–6

Cabrera AL, Willink A (1973) Biogeografía de América Latina. Programa Regional de Desarrollo Científico y Tecnológico, Departamento de Asuntos Científicos. Monografía No. 13, Serie de Biología). Washington D.C., Secretaría General de la Organización de los Estados Americanos (OEA)

Castiñeira C (2008) Aspectos de la colonización humana prehistórica del noroeste de Uruguay. Unpublished PhD Thesis, Departamento de Posgrado de la Facultad de Filosofía y Letras, Universidad de Buenos Aires, Buenos Aires

Cavallotto J, Violante R, Colombo F (2005) Evolución y cambios ambientales de la llanura costera de la cabecera del Río de la Plata. Revista de la Asociación Geológica Argentina 60(2):353–367

Dewis J, Freitas F (1970) Métodos físicos y químicos de análisis de suelos y aguas. Boletín sobre suelos FAO 10:252

Días A (1994) Repensando a Tradição Umbu através de um estudo de caso. Pontifícia Universidade Católica do Rio Grande do Sul, Porto Alegre, Tesis de Maestría

Días A (2012) Hunter gatherers occupation of south Brazil Atlantic forest: paleoenvironment and archaelogy. Quatern Int 256:12–18

Dillehay T (2000) The settlement of the Americas. A new prehistory. Basic Books, New York

del Puerto L, García-Rodríguez F, Bracco R, Castiñeira C, Blasi A, Inda H, Mazzeo N, Rodríguez A (2011) Evolución Climática Holocénica Para el Sudeste del Uruguay: Análisis Multi-proxy en Testigos de Lagunas Costeras. In: García-Rodríguez F (ed) El Holoceno en la Zona Costera de Uruguay. Universidad de la República, Montevideo, pp 117–154

Favier Dubois CM (2006) Dinámica fluvial, paleoambientes y ocupaciones humanas en la localidad arqueológica Paso Otero, Río Quequén Grande, Provincia de Buenos Aires. Intersecciones en Antropología 7:109–127

Folk R, Ward W (1957) Brazos river bar a study in the significance of grain-size parameters. J Sed Petr 27(1):3–27

Gascue A, López Mazz J, Villarmarzo E, De León V, Sotelo M, Alzugaray S (2009) La organización de la tecnología lítica de los pobladores Tempranos del este de Uruguay. Intersecciones en Antropología, Facultad de Ciencias Sociales, UNCPBA, Argentina 10:63–73

Gornitz V (2009) Ancient cultures and climate change. In: Gornitz V (ed) Enciclopedia of paleoclimatology and ancient environments. Springer, The Netherlands, pp 6–10

Goso H (1965) El Cenozoico en Uruguay. Edición Mineralógica de trabajos geológicos realizados estimados de interés general. Ministerio de Industria y Comercio. Instituto Geológico del Uruguay

Goso H (1985) El Cuaternario en el Uruguay. Estratigrafía y geología histórica. Revista de la Sociedad Uruguaya de Geología, I, vols 2–3. Montevideo, p 19

Guidon N (1989) Misión de Rescate Arqueológico, Salto Grande. Tomo I. Ministerio de Educación y Cultura, Montevideo

Hasui Y, Carneiro CDR, Coimbra AM (1975) The Ribeira folded belt. Revista Brasileira de Geociencias 5(4):257–266

Heiri O, Lotter A, Lemcke G (2001) Loss on ignition as a method for estimating organic and carbonate content in sediments: reproducibility and comparability of results. J Paleolimnol 25:101–110

Inda H (2011) Paleolimnología de cuerpos de agua someros del sudeste del Uruguay: evolución holocena e impacto humano. Editorial Académica Española, Madrid

Iriarte J (2006) Vegetation and climate change since 14,810 ^{14}C yr B.P. in southeastern Uruguay and implications for the rise of early formative societies. Quatern Res 65:44–56

Iriondo M, García J (1993) Climatic variations in the Argentine plains during the last 18,000 years. Palaeogeogr Palaeoclimatol Palaeoecol 101:209–220

López Mazz J (2001) Las estructuras tumulares (cerritos) del litoral atlántico uruguayo. Latin Amer Antiquity SAA Soc Amer Archaeol 12(3):231–255

López Mazz J, Gianotti C (2001) Diseño de proyecto y primeros resultados, de las investigaciones realizadas en la localidad arqueológica Rincón de los Indios. In: Arqueología hacia el fin del milenio, Editorial Gráficos del Sur, pp 163–174

López Mazz J (2013) Early human occupation of Uruguay: radiocarbon database and archaeological implications. Quatern Int 30:1–10

López Mazz J, Nadal O, Suárez X, De León V, Salvo X (2009) La gestión regional de los recursos minerales en las tierras bajas del este: el espacio como variable de la producción lítica. In: López Mazz J, Gascue A (eds) Arqueología Prehistórica Uruguay en el siglo XXI. Biblioteca Nacional, Montevideo, pp 39–65

López Mazz J, Aguirrezábal R, Machado A, Sotelo M, Carusso L (2010) Ocupaciones prehistóricas en el litoral atlántico uruguayo. El caso Laguna Negra. XVII Congreso Nacional de Arqueología Argentina, Mendoza

Meneghin U (2004) Urupez. Primer registro radiocarbónico (C 14) para un yacimiento con puntas líticas pisiformes del Uruguay. Orígenes 2:1–30

Meneghin U (2006) Un nuevo registro radiocarbónico (C 14) en el yacimiento Urupez II, Maldonado, Uruguay. Orígenes 5:1–7

Miotti L (2006) La fachada atlántica, como puerta de ingreso alternativa de la colonización humana de América del Sur durante la transición Pleistoceno/Holoceno. 2° Simposio Internacional del Hombre Temprano en América, Mexico DF, pp 155–188

MGAP/DGRNR/CONEAT (1979) Grupos de Suelos e Índices de Productividad. MGAP, Montevideo

Montaña J, Bossi J (1995) Geomorfología de los Humedales de la Cuenca de la Laguna Merín en el Departamento de Rocha. Facultad de Agronomía, Montevideo

Morrone JJ (2001) Biogeografía de América Latina y el Caribe. M&T–Manuales & Tesis. SEA, vol 3. Zaragoza, 148 pp

Morrone JJ (2006) Biogeographic areas and transition zones of Latin America and the Caribbean Islands based on panbiogeographic and cladistic analyses of the entomofauna. Annu Rev Entomol 51:467–494

Munsell Soil Color Charts (1992) Revised edition. Macbeth, Division of Kollmorgen Instruments Corp., New York

Nami H (2013) Archaeology, paleoindian research and Lithic technology in the middle Negro River, central Uruguay. Archaeol Discov 1:1–22

http://www.scirp.org/journal/PaperInformation.aspx?PaperID=35118

Nicholson S, Flohn H (1980) African environmental and climatic changes and the general atmospheric circulation in late Pleistocene and Holocene, Springer. Clim Change 2:313

Panario D, Gutiérrez O (1999) The continental uruguayan Cenozoic: an overview. Quatern Int 62:75–84

Panario D, Gutiérrez O (2011) Introducción a la Geomorfología de Lagunas Costeras, Lagos Someros y Charcas de Uruguay. In: García-Rodríguez F (ed) El Holoceno en la Zona Costera de Uruguay. Universidad de la República, Montevideo, pp 49–63

Piñeiro G, Suárez R (2012) Paleosuelos en el Holoceno del Río Cuareim. III Jornadas del Cenozoico de la Sociedad Uruguaya de Geología, Montevideo, pp 1–5

Piñeiro G, Suárez R, Gascue A (1999) Transición Pleistoceno-Holoceno en los Sitios Arqueológicos del Río Uruguay medio: Sedimentos. I Jornada del Cenozoico. Facultad de Ciencias, Montevideo

Piovano E, Ariztegui D, Córdoba F, Cioccale M, Sylvestre F (2009) Hydrological variability in South America below the Tropic of Capricorn (Pampas and eastern Patagonia, Argentina) during the last 13.0 ka. Past climate variability from the Last Glacial Maximum to the Holocene in South America and surrounding regions (Focus on local and large scale teleconnexions), pp 1–42

Politis G (1991) Fishtail projectile points in the southern cone of south America: an overview. In: Bonnichsen R, Turnmire K (eds) Clovis, origins and adaptations. Center for the study of the First Americans, College Station, TX ASM University, pp 287–301

Preciozzi F, Spoturno J, Heinzen W, Rossi P (1985) Carta Geológica del Uruguay a Escala 1:500.000. Ministerio de Industria y Eenergía. Dirección Nacional de Minería y Geología. Montevideo

Prohaska F (1976) The climate of Argentina, Paraguay and Uruguay. In: Schwerdtfeger E (ed) Climate of central and south America. World survey of climatology. Elsevier, Amsterdam, pp 57–69

Roberto Bracco, Laura del Puerto, Hugo Inda, Daniel Panario, Carola Castiñeira, Felipe García-Rodríguez, (2011) The relationship between emergence of mound builders in SE Uruguay and climate change inferred from opal phytolith records. Quaternary International 245 (1):62-73

Sombroek W, Averbek H, Duran A (1969) Soil Studies in the Merim Lagoon Basin. Merim Lagoon Regional Proyect. Treinta y Tres, Uruguay

Sotelo M, López Mazz J (2015) Recuperación de restos por flotación y zaranda de agua. Componentes tempranos del sitio arqueológico Rincón de los Indios. VI Jornadas de Investigación de la Facultad de Humanidades y Ciencias de la Educación, Montevideo

Suárez R, López Mazz J (2003) Archaeology of the Pleistocene-Holocene transition in Uruguay: an overview. Quatern Int 109–110:65–76

Suárez R (2011) Movilidad, acceso y uso de ágata translúcida por los cazadores-recolectores tempranos durante la transición Pleistoceno-Holoceno en el norte de Uruguay (ca. 11,000-8500 A.P.). Latin Amer Antiquity 22(3):359–383

Ubilla M (1999) Dataciones radiocarbónicas ^{14}C para la Formación Dolores (Río Santa Lucía Departamento de Canelones y Arroyo Gutiérrez Chico, Departamento de Río Negro) y comentario sobre la fauna de vertebrados asociada. Revista de la Sociedad Uruguaya de Geología III(6):48–54

United States Department of Agriculture (2006) Soil taxonomy a basic system of soil classification for making and interpreting soil surveys

Vega J (2007) Caracterización de eventos climáticos del Pleistoceno superior-Holoceno mediante el estudio sedimentológico de la Laguna Grande (Sierra de Neila, NO sistema ibérico). Sociedad Geológica de España 20:1–2

Wentworth C (1922) A scale of grade and class terms for clastic sediments. J Geol 30(5):377–392

Author Index

H. Inda Ferrero and F. García Rodríguez (eds.), *Advances in Coastal Geoarchaeology in Latin America*, The Latin American Studies Book Series,
https://doi.org/10.1007/978-3-030-17828-4

The manufacturer's authorised representative in the EU is Springer Nature Customer Service Centre GmbH, Europaplatz 3, 69115 Heidelberg, Germany. If you have any concerns regarding our products, please contact ProductSafety@springernature.com

Printed and bound by CPI Group (UK) Ltd, Croydon, CR0 4YY
30/09/2026
02225509-0002